Springer **M**onographs in **M**athematics

More information about this series at http://www.springer.com/series/3733

John Rhodes • Pedro V. Silva

Boolean Representations of Simplicial Complexes and Matroids

 Springer

John Rhodes
Department of Mathematics
University of California at Berkeley
Berkeley, CA, USA

Pedro V. Silva
Department of Mathematics
University of Porto
Porto, Portugal

ISSN 1439-7382   ISSN 2196-9922   (electronic)
Springer Monographs in Mathematics
ISBN 978-3-319-38367-5   ISBN 978-3-319-15114-4   (eBook)
DOI 10.1007/978-3-319-15114-4

Mathematics Subject Classification (2010): 03E05, 03G10, 05-00, 05B25, 05B30, 05B35, 05D15, 05E45, 06-00, 06A11, 06A12, 06B15, 15B34, 16Y60, 51E14, 55P15, 55U10

Springer Cham Heidelberg New York Dordrecht London
© Springer International Publishing Switzerland 2015
Softcover reprint of the hardcover 1st edition 2015

Printed on acid-free paper

Springer International Publishing AG Switzerland is part of Springer Science+Business Media (www.springer.com)

In memory of Gian-Carlo Rota

Acknowledgments

Both the authors thank Anne Schilling, Benjamin Steinberg, James Oxley, and Stuart Margolis for their valuable comments.

John Rhodes would like to thank his remarkable wife, Laura Morland, for all her love and support.

Pedro V. Silva is grateful to his wife Margarida for her enduring support and for everything else! He also acknowledges support from

- The European Regional Development Fund through the programme COMPETE and the Portuguese Government through FCT (Fundação para a Ciência e a Tecnologia) under the project PEst-C/MAT/UI0144/2013
- CNPq (Brazil) through a BJT-A grant (process 313768/2013-7)

Contents

Chapter 1
Introduction

Matroids were introduced by Whitney in 1935 [55] to generalize to an abstract setting the concept of independence originated in linear algebra. One of the strongest features of matroid theory is its rich geometric theory, but the classical representation theory of matroids is not entirely successful: it is well known that not all matroids admit a field representation [39]. Attempts have been made to replace fields by more general structures such as *partial fields* [49] or *quasi fields* [18], but they still failed to cover all matroids.

The main goal of this monograph is to propose a new representation theory in a generalized context going beyond matroids, where the latter become representable in all cases, and still rich enough to allow geometric, topological and combinatorial applications.

Throughout the text, we shall give evidence of the geometric potential of these new ideas. They extend in many aspects the known geometric theory for matroids, but they also raise new perspectives in the matroid world. In particular, we believe that our results and techniques may be of interest in connection with several of the famous conjectures and constructions for matroids. See Chap. 9 on open questions for details, particularly Questions 9.1.4 and 9.1.5. We note also that our theory extends to finite posets, see Sects. 3.2 and 9.5.

Matroids are of course particular cases of (abstract) *simplicial complexes* (also known as *hereditary collections*) $\mathcal{H} = (V, H)$. In the topological (respectively combinatorial) terminology,

- Elements of H are called *simplexes* (respectively *independent*);
- Maximal elements of H, with respect to inclusion, are called *facets* (respectively *bases*);
- The *dimension* of a facet X is $|X| - 1$ (the *rank* of a basis X is $|X|$).

We adopt the topological terminology in this text.

© Springer International Publishing Switzerland 2015
J. Rhodes, P.V. Silva, *Boolean Representations of Simplicial Complexes and Matroids*, Springer Monographs in Mathematics,
DOI 10.1007/978-3-319-15114-4_1

Matroids are defined through the axiom known as the *exchange property* (EP), and a weaker condition we consider is the *point replacement property* (PR) (see page 32). However, (PR) appears to be too weak to induce a rich geometric or representation theory, so we are in fact proposing an alternative third axiom which will be fully explained in this text and informally explained below in this Introduction,

(BR) $\mathcal{H}$ is boolean representable over the superboolean semiring $\mathbb{SB}$,

an axiom strictly stronger than (PR) (Proposition 5.1.2 and Example 5.2.12) but strictly weaker than (EP) (Theorem 5.2.10 and Example 5.2.11(iii)). Thus all matroids satisfy (BR) and so results on (BR) apply to matroids. See page 7 for further discussion of the axioms.

Therefore the reader may have two viewpoints on these new ideas:

• To consider them as a source of new concepts, techniques and problems for matroid theory;
• To consider the new class of boolean representable simplicial complexes as a brave new world to explore (we believe many of the theorems in matroid theory will extend to boolean representable simplicial complexes).

We have tried to suit both perspectives.

In 2006 [25] (see also [31]), Zur Izhakian had the seminal idea of considering independence for columns of a boolean matrix, arising in the context of (super)tropical semirings (tropical geometry). In 2008, due to the prevalence of boolean matrices throughout mathematics, the first author saw massive applications of this idea to various areas, more specifically combinatorial geometry and topology, combinatorics and algebra. Many of the results go through for infinite posets, lattices and boolean matrices, but we limit our exposition here to the finite case due to length considerations.

The foundations of the theory were formulated and developed by Izhakian and the first author in [28–30]. Subsequently, the theory was matured and developed by the present authors in this monograph and the paper [42], devoted to applications to graph theory.

The well-known *boolean semiring* $\mathbb{B}$ can be built as the quotient of the semiring $(\mathbb{N}, +, \cdot)$ (for the standard operations) by the congruence which identifies all $n \geq 1$. The *superboolean semiring* $\mathbb{SB}$ is the quotient of the semiring $(\mathbb{N}, +, \cdot)$ by the congruence which identifies all $n \geq 2$, having therefore 3 elements: 0, 1 and "at least 2". So the world has successively witnessed the plain $1 + 1 = 2$, Galois's $1 + 1 = 0$, Boole's $1 + 1 = 1$ and now finally $1 + 1 = $ at least 2!

We call $\mathcal{G} = \{0, 2\}$ the *ghost ideal* of $\mathbb{SB}$. According to the concept of independence for supertropical semirings adopted by Izhakian and Rowen [25, 31, 33], n vectors $C_1, \ldots, C_n \in \mathbb{SB}^m$ are *independent* if

$$\lambda_1 C_1 + \ldots + \lambda_n C_n \in \mathcal{G}^m \quad \text{implies} \quad \lambda_1 = \ldots = \lambda_n = 0 \qquad (1.1)$$

for all $\lambda_1, \ldots, \lambda_n \in \{0, 1\}$. As we show in Proposition 2.2.5, this is equivalent to saying that the corresponding $m \times n$ matrix has a square $n \times n$ submatrix M congruent to some lower unitriangular matrix, i.e. by independently permuting rows/columns of M, one can get a matrix of the form

$$\begin{pmatrix} 1 & 0 & 0 & \ldots & 0 \\ ? & 1 & 0 & \ldots & 0 \\ ? & ? & 1 & \ldots & 0 \\ \vdots & \vdots & \vdots & \ddots & \vdots \\ ? & ? & ? & \ldots & 1 \end{pmatrix} \tag{1.2}$$

Moreover, M satisfies the above property if and only if M has permanent 1, where *permanent* denotes the positive version of the determinant (and therefore suitable for operations on semirings, and in particular for the $\mathbb{SB}$ semiring).

Then the *rank* of a matrix, i.e. the maximum number of independent columns, turns out to be the maximum size of a square submatrix with permanent 1. Notice the similarity to the classical situation in matroid theory, where the matrix has coefficients in a field F and permanent 1 is replaced by determinant $\neq 0$. Note also that in the field case the number of rows can always be chosen to be the rank of the matroid, but not in the boolean case (see Proposition 5.7.12 for an example).

We note that the notion of rank of a matrix we adopt is one of a number in the literature. For alternative notons, the reader is referred to [15].

Following Izhakian and Rhodes [28], we can define the class $\mathcal{BR}$ of *boolean representable simplicial complexes* as the set of all simplicial complexes $\mathcal{H} = (V, H)$ for which there exists, for some $n \geq 1$, an $n \times |V|$ boolean matrix M such that, for every $X \subseteq V$, we have $X \in H$ if and only if the columns of M corresponding to the elements of X are independent as defined in (1.1). This is equivalent to saying that M admits, for some $R \subseteq \{1, \ldots, n\}$, a square submatrix $M[R, X]$ congruent to (1.2), i.e. with permanent 1.

As a matter of fact, Izhakian and Rhodes proved in [28] that every simplicial complex $\mathcal{H} = (V, H)$ admits a representation over $\mathbb{SB}$ (the entries may be 0, 1 or 2). This is of course too general a class to allow the development of an interesting geometric theory in matroid style, but the restriction to boolean matrices proved to be a much more interesting bet.

The relationship between boolean matrices and finite lattices is one of the cornerstones of this monograph, which we explain now.

Let $M = (a_{ij})$ be an $m \times n$ boolean matrix. Write $C = \{1, \ldots, n\}$. For $i = 1, \ldots, m$, let

$$Z_i = \{j \in C \mid a_{ij} = 0\}$$

be the set of positions of the 0's in the ith row (why we take the 0's instead of the usual 1's will be explained below). We define

$$\mathrm{Fl}\, M = \{\cap_{i \in D} Z_i \mid D \subseteq C\}.$$

Since $\mathrm{Fl}\, M$ is closed under intersection, it is a $\wedge$-semilattice, and becomes then a (finite) lattice with the determined join (see (3.4)), termed the *lattice of flats* of M. This terminology is inspired by the lattice of flats of a matroid, which happens to be an important particular case of $\mathrm{Fl}\, M$ as we shall se later on.

For $j \in C$, define

$$Y_j = \cap\{Z_i \mid a_{ij} = 0\},$$

which may be viewed as the closure of $\{j\}$. If we assume that M has no zero columns, it turns out (Proposition 3.4.1) that $\mathrm{Fl}\, M$ is $\vee$-generated by the subset

$$\mathcal{Y}(M) = \{Y_1, \ldots, Y_n\}.$$

Conversely, if L is a finite lattice $\vee$-generated by A (so that we can assume that the bottom element B is *not* in A), we define the boolean representation of (L, A) to be the matrix $M(L, A) = (m_{xa})$ defined by

$$m_{xa} = \begin{cases} 0 & \text{if } x \geq a \\ 1 & \text{otherwise} \end{cases}$$

for all $x \in L$ and $a \in A$. The reasons for this placement of 0 and 1 are explained below.

We show in Sect. 3.5 that under mild assumptions the operators $M \mapsto (\mathrm{Fl}\, M, \mathcal{Y}(M))$ and $(L, A) \mapsto M(L, A)$ are mutually inverse, therefore we may view boolean matrices and finite $\vee$-generated lattices as alternative perspectives of the same object. We now wish to see what boolean independent X columns of $M(L, A)$ correpond to in (L, A), and conversely.

Let L be a finite lattice $\vee$-generated by A. We define c-independence in (L, A) in terms of independence for the corresponding vector columns of $M(L, A)$. In Proposition 3.6.2, we show that $X \subseteq A$ is c-independent as a subset of L if and only if X admits an enumeration $x_1, \ldots, x_k$ such that

$$(x_1 \vee \ldots \vee x_k) > (x_2 \vee \ldots \vee x_k) > \ldots > (x_{k-1} \vee x_k) > x_k. \tag{1.3}$$

Notice that, for any lattice L, we may set $A = L \setminus \{B\}$ and use (1.3) to get a notion of independence for L. This idea has been used for geometric lattices, and we extend it here to arbitrary finite lattices.

Furthermore, using the closure operator Cl_L on $(2^A, \subseteq)$ induced by L, we show also that c-independence (and therefore (1.3)) are also equivalent to X being a transversal of the successive differences for some chain of $\text{Fl}\, M(L, A)$. See below (3.10) and Proposition 3.6.2 for details.

We prove also (Proposition 3.6.4) that the rank of $M(L, A)$ equals the height of the lattice L.

Now the relationship between boolean matrices and finite lattices opens new perspectives on boolean representability. Note that independence and rank can be understood omiting any reference to $\mathbb{SB}$ and can be considered as purely combinatorial properties of boolean matrices and finite lattices.

The concept of flat, defined for an arbitrary simplicial complex, plays a major role in boolean representation theory. In matroid theory, there are plenty of equivalent definitions, but they are not necessarily equivalent for an arbitrary complex. We choose the generalization of the definition which uses the independent sets in the matroid setting: given a simplicial complex $\mathcal{H} = (V, H)$ and $X \subseteq V$, we say that X is a *flat* if

$$\forall I \in H \cap 2^X \ \forall p \in V \setminus X \quad I \cup \{p\} \in H.$$

It is immediate that the intersection of flats is still a flat. Hence the flats of $\mathcal{H}$ constitute a lattice under the determined join, denoted by $\text{Fl}\, \mathcal{H}$.

Let M be a boolean matrix representation of a simplicial complex $\mathcal{H} = (V, H)$ (so M has column space V). We often assume that all columns are nonzero and distinct so that $\mathcal{H}$ is *simple* (all sets of two or less elements are independent). In each row of M, the positions of the zeroes define a flat of $\mathcal{H}$, so by closing under all intersections $\text{Fl}\, M$ constitutes a $\wedge$-subsemilattice of $\text{Fl}\, \mathcal{H}$.

Now the $(\text{Fl}\, \mathcal{H}) \times V$ boolean representation $\text{Mat}\, \mathcal{H} = (m_{Xp})$ defined by

$$m_{Xp} = \begin{cases} 0 & \text{if } p \in X \\ 1 & \text{otherwise} \end{cases}$$

is the canonical ("biggest") boolean representation of $\mathcal{H}$ (if $\mathcal{H}$ is boolean representable) in some precise sense (see Theorem 5.2.5). Note that the roles of 0 and 1 are reversed with respect to the standard representation practice. Due to the notation $0^c = 1$, $1^c = 0$, we have adopted the terminology c-*independence* and c-*rank* in this sense to avoid any possible confusion.

But $\text{Mat}\, \mathcal{H}$ is far from being the most economical representation, and smaller matrices M lead to smaller lattices $\text{Fl}\, M$ representing $\mathcal{H}$, even in the matroid case. Thus we establish the concepts of *lattice representation* and *lattice of all boolean representations* of a (boolean representable) simplicial complex, which includes all matroids. This provides a representation theory comprised of all boolean repesentations of $\mathcal{H}$. The role played by lattices in the whole theory explains why we need to develop a theory of boolean representations of $\vee$-generated lattices prior to engaging on the representation of simplicial complexes. In fact this theory can be extended to arbitrary (finite) posets, with the help of the Dedekind-MacNeille completion.

Each lattice representation (L, V) of $\mathcal{H} = (V, H)$ induces a closure operator Cl_L on 2^V, and using the representation theory developed for lattices, we get the following (see Theorem 5.4.2): for every $X \subseteq V$, we have $X \in H$ if and only if X admits an enumeration $x_1, \ldots, x_k$ such that

$$\text{Cl}_L(x_1, \ldots, x_k) \supset \text{Cl}_L(x_2, \ldots, x_k) \supset \ldots \supset \text{Cl}_L(x_k) \supset \text{Cl}_L(\emptyset).$$

This equivalence is of course valid when we take $L = \text{Fl } \mathcal{H}$. Then $\text{Cl}_L(X) = \overline{X}$ denotes the smallest flat containing X for every $X \subseteq V$.

The advantage of lattice representations over matrix representations for a given complex $\mathcal{H}$ is that they can be quasi-ordered using the concept of $\vee$-*map*, which in our context replaces the strong/weak maps from matroid theory. Adding an extra element to become the bottom and identifying isomorphic $\vee$-generated lattices, the lattice representations of $\mathcal{H}$ become a lattice of their own, denoted by $\text{LR}_0 \, \mathcal{H}$ (Theorem 5.5.5). The top element is the canonical representation by the lattice of flats $\text{Fl } \mathcal{H}$.

Understanding the structure of this lattice is important, identifying in particular the *atoms* (which are the *minimal* lattice representations) and the *sji* (strictly join irreducible) elements. The join operator is "stacking matrices", see Corollary 5.5.8, and every boolean representation is the stack of some sji representations. Together with computing the minimum degree (number of rows) of a matrix representation (*mindeg*), these constitute most challenging problems for a (boolean representable) simplicial complex, new even in the matroid case. In Sect. 5.7, we perform these computations for some interesting particular cases which include the tetrahedron matroid T_3, the Fano matroid F_7 and the uniform matroids $U_{3,n}$ for $n \geq 5$. The next task is to perform these calculations in more sophisticated matroid examples, see Sect. 9.1 and Question 9.1.3.

In more general terms, progress has been achieved in the case of paving simplicial complexes, namely in the case of low dimensions. A simplicial complex $\mathcal{H} = (V, H)$ of dimension d is called *paving* if H contains every d-subset of V.

We focus our attention on the class $\text{BPav}(2)$ of boolean representable paving simplicial complexes of dimension 2. Indeed, we develop geometric tools for computing the flats in $\text{BPav}(2)$, giving insight into the boolean representation theory. We propose two approaches:

The first approach involves the concept of *partial Euclidean geometry* (PEG), an (abstract) system of points and lines where each line has at least two points and intersects any other line in at most one point. Given a matrix representation M of $\mathcal{H} = (V, H) \in \text{BPav}(2)$, we can build a PEG $\text{Geo } M = (V, \mathcal{L}_M)$ as follows: for each nonzero row having at least 2 zeroes, we take the corresponding flat of M to be a line. Using the concept of *potential line* (a set with at least two points whose addition to a PEG results in a PEG, see page 89) we can compute both H and $\text{Fl } \mathcal{H}$ from $\text{Geo } M$ (Lemma 6.3.3 and Theorem 6.3.4). We note that this approach may be generalized to paving simplicial complexes of higher dimensions, considering also geometries of arbitrary dimension. A hint of this is given in Sect. 5.7.4, devoted to *Steiner systems*. See also the open questions in Sect. 9.3.

The second approach is graph-theoretic and relies on the definition of a graph ΓM to pursue similar objectives (Theorem 6.3.6 and Proposition 6.3.7). Concepts such as *anticliques* and *superanticliques* play a key role in this approach.

The particular case of the canonical representation $M = \text{Mat } \mathcal{H}$ is of utmost importance: we go deeply into the study of the *graph of flats* $\Gamma\text{Fl } \mathcal{H} = \Gamma \text{ Mat } \mathcal{H} = (V, E)$ defined by

$$p \text{ --- } q \text{ is an edge in } E \text{ if and only if } \overline{pq} \subset V$$

for all $p, q \in V$ distinct (where $\overline{pq}$ is the smallest flat containing pq).

The graph of flats plays a major role in the topological applications of the theory (homotopy type of the simplicial complex $\mathcal{H}$). We can also compute mindeg $\mathcal{H}$ if $\Gamma\text{Fl } \mathcal{H}$ is disconnected (Theorem 6.5.1).

One of the main topological applications of our theory is the determination of the homotopy type making use of the concept of *shellability* for non pure complexes, introduced by Björner and Wachs [5, 6]. They prove that the existence of a *shelling* for a simplicial complex $\mathcal{H}$ (an enumeration of the facets of $\mathcal{H}$ satisfying favorable conditions) implies that the geometric realization $\| \mathcal{H} \|$ of $\mathcal{H}$ is homotopically equivalent to a *wedge of spheres*, and the *Betti numbers* are easy to compute.

Using the graph of flats $\Gamma\text{Fl } \mathcal{H}$, we succeed on identifying the shellable complexes $\mathcal{H} \in \text{BPav}(2)$: they are precisely those complexes such that $\Gamma\text{Fl } \mathcal{H}$ contains at most two connected components or contains exactly one nontrivial connected component (Theorem 7.2.8). These are also the sequentially Cohen-Macaulay complexes of BPav(2) (Corollary 7.2.9). We also prove that every finite graph is isomorphic to the graph of flats of some $\mathcal{H} \in \text{Pav}(2)$, except in the case of a disjoint union $K_r \sqcup K_s \sqcup K_1$ of complete graphs with $r, s > 1$ (Theorem 7.3.1).

The class $\mathcal{BR}$ of boolean representable simplicial complexes is not closed under the most common operators, except for *restriction* and *isomorphism*, see Chap. 8. Those who seek closure under contraction and dual must restrict to matroids and use the representation theory for a fixed matroid. As it turns out, if all the contractions of a simplicial complex satisfy (PR), it must be a matroid (Proposition 8.3.6). Moreover, every simplicial complex is the contraction of a boolean representable simplicial complex (Proposition 8.3.7). Thus the concepts of *minor* and *minor-closed subclass*, so important in the contexts of graphs (Robertson-Seymour Theorem [16, Chapter 12]) and matroids (see [21]), cannot be directly applied in our generalized context. However, we can get away with restriction and isomorphism only, introducing the concept of *prevariety* of simplicial complexes: a class of simplicial complexes closed under restriction and isomorphism.

A prevariety is *finitely based* if it can be defined through a finite set Σ of forbidden restrictions (*basis*). Bounding the dimension of the complexes in the prevariety is important to get finitely based, so if $\mathcal{V}$ is a prevariety, we denote by

$\mathcal{V}_d$ the prevariety formed by the complexes in $\mathcal{V}$ with dimension $\leq d$. We can prove that $\mathcal{V}_d$ is finitely based for the most natural prevarieties of simplicial complexes.

The maximum number of vertices of a complex in Σ is the *size* of the basis Σ and the size of a prevariety $\mathcal{V}$, denoted by siz $\mathcal{V}$, is the minimum size of such a basis. Among other results, we show that siz $\mathcal{PB}_d = (d + 1)(d + 2)$ for every $d \geq 2$, where $\mathcal{PB}$ denotes the prevariety of boolean representable paving simplicial complexes (Theorem 8.5.2(ii)). We also show that siz $\mathcal{BR}_d \leq (d + 1)^2 d^{2d} + d + 1$ (Theorem 8.5.4(iii)).

Part of the material contained in this monograph (and other things as well) can be found under a slightly different perspective in our arXiv preprint [43]. We have sought to extend to this new boolean setting many of the results found in Stanley's monograph [52], and in later work to include Möbius functions, see [45].

We should remark that boolean representable simplicial complexes are just one of the natural ways of generalizing matroids. Another natural generalization, built over a different property, leads to the concept of *greedoid* (see the survey by Björner and Ziegler [8]). In the case of boolean representable simplicial complexes, we have the means to characterize independence through chains in a lattice (which may be assumed to be the lattice of flats), similarly to matroids; in the case of greedoids, the exchange property of matroids is kept but hereditarity is not required (so a greedoid is a simplicial complex if and only if it is a matroid). It turns out that in both cases matroids can be viewed as the *commutative case*, and one of the topics of our near future research is to establish all the relationships (of algebraic, combinatorial, geometric and syntactic nature) between boolean representable simplicial complexes, matroids, greedoids and the important subclass of *interval greedoids*.

Chapter 2
Boolean and Superboolean Matrices

We introduce in this chapter the superboolean semiring $\mathbb{SB}$ and the core of the theory of (boolean) matrices over $\mathbb{SB}$, with special emphasis on the concepts of independence of vectors and rank. These matrices are used to represent various kinds of algebraic and combinatorial objects, namely posets and simplicial complexes, especially matroids.

2.1 The Boolean and the Superboolean Semirings

A *commutative semiring* is an algebra $(S, +, \cdot, 0, 1)$ of type $(2, 2, 0, 0)$ satisfying the following properties:

(CS1) $(S, +, 0)$ and $(S, \cdot, 1)$ are commutative monoids;
(CS2) $a \cdot (b + c) = (a \cdot b) + (a \cdot c)$ for all $a, b, c \in S$;
(CS3) $a \cdot 0 = 0$ for every $a \in S$.

To avoid trivial cases, we assume that $1 \neq 0$. If the operations are implicit, we denote this semiring simply by S.

If we only require commutativity for the monoid $(S, +, 0)$ and use both left-right versions of (CS2) and (CS3), we have the general concept of *semiring*.

Clearly, commutative semirings constitute a variety of algebras of type $(2, 2, 0, 0)$, and so universal algebra provides the concepts of congruence, homomorphism and subsemiring. In particular, an equivalence relation σ on S is said to be a *congruence* if

$$(a\sigma b \,\wedge\, a'\sigma b') \;\Rightarrow\; ((a + a')\sigma(b + b') \,\wedge\, (a \cdot a')\sigma(b \cdot b')).$$

© Springer International Publishing Switzerland 2015
J. Rhodes, P.V. Silva, *Boolean Representations of Simplicial Complexes and Matroids*, Springer Monographs in Mathematics,
DOI 10.1007/978-3-319-15114-4_2

In this case, we get induced operations on $S/\sigma = \{a\sigma \mid a \in S\}$ through

$$a\sigma + b\sigma = (a+b)\sigma, \quad a\sigma \cdot b\sigma = (a \cdot b)\sigma.$$

If σ is not the universal relation, then $1\sigma \neq 0\sigma$ and so $(S/\sigma, +, \cdot, 0\sigma, 1\sigma)$ is also a commutative semiring, the *quotient* of S by σ.

The natural numbers $\mathbb{N} = \{0, 1, 2, \ldots\}$ under the usual addition and multiplication provide a most important example of a commutative semiring. For every $m \in \mathbb{N}$, we define a relation σ_m on $\mathbb{N}$ through

$$a\sigma_m b \quad \text{if } a = b \text{ or } a, b \geq m.$$

Then σ_m is a congruence on $\mathbb{N}$ and

$$n\sigma_m = \begin{cases} \{n\} & \text{if } n < m \\ \{m, m+1, \ldots\} & \text{otherwise} \end{cases}$$

Hence $\{0, \ldots, m\}$ is a cross-section for $\mathbb{N}/\sigma_m$.

We can define the *boolean semiring* as the quotient

$$\mathbb{B} = \mathbb{N}/\sigma_1.$$

As usual, we view the elements of $\mathbb{B}$ as the elements of the cross-section $\{0, 1\}$. Addition and multiplication are then described respectively by

+	0	1		·	0	1
0	0	1		0	0	0
1	1	1		1	0	1

Similarly, we can define the *superboolean semiring* as the quotient

$$\mathbb{SB} = \mathbb{N}/\sigma_2.$$

We can view the elements of $\mathbb{SB}$ as the elements of the cross-section $\{0, 1, 2\}$. Addition and multiplication are then described respectively by

+	0	1	2		·	0	1	2
0	0	1	2		0	0	0	0
1	1	2	2		1	0	1	2
2	2	2	2		2	0	2	2

Since $1 + 1$ takes different values in both semirings, it follows that $\mathbb{B}$ is not a subsemiring of $\mathbb{SB}$. However, $\mathbb{B}$ is a homomorphic image of $\mathbb{SB}$ (through the canonical mapping $n\sigma_2 \mapsto n\sigma_1$).

For an alternative perspective of $\mathbb{SB}$ as a supertropical semiring, the reader is referred to Sect. A.1 of the Appendix.

2.2 Superboolean Matrices

Given a semiring S, we denote by $\mathcal{M}_{m \times n}(S)$ the set of all $m \times n$ matrices with entries in S. We write also $\mathcal{M}_n(S) = \mathcal{M}_{n \times n}(S)$. Addition and multiplication are defined as usual.

Given $M = (a_{ij}) \in \mathcal{M}_{m \times n}(S)$ and nonempty $I \subseteq \{1, \ldots, m\}$, $J \subseteq \{1, \ldots, n\}$, we denote by $M[I, J]$ the submatrix of M with entries a_{ij} ($i \in I$, $j \in J$). For all $i \in \{1, \ldots, m\}$ and $j \in \{1, \ldots, n\}$, we write also

$$M[\bar{i}, \bar{j}] = M[\{1, \ldots, m\} \setminus \{i\}, \{1, \ldots, n\} \setminus \{j\}].$$

Finally, we denote by $M[i, \_]$ the ith row vector of M, and by $M[\_, j]$ the jth column vector of M.

The results we present in this section are valid for more general semirings (any *supertropical semifield*, actually, see [25, 31] and Sect. A.1 in the Appendix), but we shall discuss only the concrete case of $\mathbb{SB}$.

Let S_n denote the symmetric group on $\{1, \ldots, n\}$. The *permanent* of a square matrix $M = (m_{ij}) \in \mathcal{M}_n(\mathbb{SB})$ (a positive version of the determinant) is defined by

$$\mathrm{Per}\, M = \sum_{\pi \in S_n} \prod_{i=1}^{n} m_{i, i\pi}.$$

Note that this formula coincides with the formula for the determinant of a square matrix over the two-element field $\mathbb{Z}_2$ (but interpreting the operations in $\mathbb{SB}$). The classical results on determinants involving only a rearrangement of the permutations extend naturally to $\mathbb{SB}$. Therefore we can state the two following propositions without proof:

Proposition 2.2.1. *Let $M = (m_{ij}) \in \mathcal{M}_n(\mathbb{SB})$ and let $p \in \{1, \ldots, n\}$. Then*

$$\mathrm{per}\, M = \sum_{j=1}^{n} m_{pj} (\mathrm{per}\, M[\bar{p}, \bar{j}]) = \sum_{i=1}^{n} m_{ip} (\mathrm{per}\, M[\bar{i}, \bar{p}]).$$

Proposition 2.2.2. *The permanent of a square superboolean matrix remains unchanged by:*

  (i) *Permuting two columns;*
 (ii) *Permuting two rows;*
(iii) *Transposition.*

Next we present definitions of independence and rank appropriate to the context of superboolean matrices, introduced by Izhakian in [25] (see also [28, 31]). For alternative notions in the context of semirings, see [15]. We need to introduce the *ghost ideal*

$$\mathcal{G} = \{0, 2\} \subseteq \mathbb{SB}$$

(see Sect. A.1 for more details on ghost ideals).

Let $\mathbb{SB}^n$ denote the set of all vectors $V = (v_1, \ldots, v_n)$ with entries in $\mathbb{SB}$. Addition and the scalar product $\mathbb{SB} \times \mathbb{SB}^n \to \mathbb{SB}^n$ are defined the obvious way.

We say that the vectors $V^{(1)}, \ldots, V^{(m)} \in \mathbb{SB}^n$ are *independent* if

$$\lambda_1 V^{(1)} + \ldots + \lambda_m V^{(m)} \in \mathcal{G}^n \quad \text{implies} \quad \lambda_1, \ldots, \lambda_m = 0$$

for all $\lambda_1, \ldots, \lambda_m \in \{0, 1\}$. Otherwise, they are said to be *dependent*. The contrapositive yields that $V^{(1)}, \ldots, V^{(m)}$ are dependent if and only if there exists some nonempty $I \subseteq \{1, \ldots, m\}$ such that $\sum_{i \in I} V^{(i)} \in \mathcal{G}^n$.

The next lemma discusses independence when we extend the vectors by one further component:

Lemma 2.2.3. *Let* $X = \{V^{(1)}, \ldots, V^{(m)}\} \subseteq \mathbb{SB}^n$ *and* $Y = \{W^{(1)}, \ldots, W^{(m)}\} \subseteq \mathbb{SB}^{n+1}$ *be such that* $V^{(i)} = (v_1^{(i)}, \ldots, v_n^{(i)})$ *and* $W^{(i)} = (a^{(i)}, v_1^{(i)}, \ldots, v_n^{(i)})$ *for* $i \in \{1, \ldots, m\}$. *Then:*

  (i) If X is independent, so is Y;
  (ii) If $a^{(1)}, \ldots, a^{(m)} \in \mathcal{G}$, then X is independent if and only if Y is independent.

Proof. (i) Assume that X is independent. Let $\lambda_1, \ldots, \lambda_m \in \{0, 1\}$ be such that $\lambda_1 W^{(1)} + \ldots + \lambda_m W^{(m)} \in \mathcal{G}^{n+1}$. Then $\lambda_1 V^{(1)} + \ldots + \lambda_m V^{(m)} \in \mathcal{G}^n$ and so $\lambda_1 = \ldots = \lambda_m = 0$ since X is independent. Thus Y is independent.

(ii) The direct implication follows from (i). Assume now that Y is independent. Let $\lambda_1, \ldots, \lambda_m \in \{0, 1\}$ be such that $\lambda_1 V^{(1)} + \ldots + \lambda_m V^{(m)} \in \mathcal{G}^n$. Since $a^{(1)}, \ldots, a^{(m)} \in \mathcal{G}$, we get $\lambda_1 W^{(1)} + \ldots + \lambda_m W^{(m)} \in \mathcal{G}^{n+1}$ and so $\lambda_1 = \ldots = \lambda_m = 0$ since Y is independent. Thus X is independent. $\square$

We start now to address independence in the context of a matrix. Two matrices M and M' are said to be *congruent* and we write $M \cong M'$ if we can transform one into the other by permuting rows and permuting columns independently. A row of a superboolean matrix is called a *marker* if it has one entry 1 and all the remaining entries are 0.

Lemma 2.2.4 ([28, Cor. 3.4]). *Let* $M \in \mathcal{M}_{m \times n}(\mathbb{SB})$ *be such that the column vectors* $M[\_, j]$ *(*$j \in \{1, \ldots, n\}$*) are independent. Then M has a marker.*

Proof. Let $M = (a_{ij})$. By independence, we must have

$$\sum_{j=1}^{n} M[\_, j] \notin \mathcal{G}^n.$$

Hence $a_{i1} + \ldots + a_{in} = 1$ for some $i \in \{1, \ldots, m\}$ and so the ith row of M is a marker. $\square$

The following result discusses independence for the row/column vectors of a square superboolean matrix. The equivalence of the three first conditions is due to Izhakian [25] (see also [31]), the remaining equivalence to Izhakian and

Rhodes [28]. Recall that a square matrix of the form

$$\begin{pmatrix} 1 & 0 & 0 & \ldots & 0 \\ ? & 1 & 0 & \ldots & 0 \\ ? & ? & 1 & \ldots & 0 \\ \vdots & \vdots & \vdots & \ddots & \vdots \\ ? & ? & ? & \ldots & 1 \end{pmatrix}$$

is called *lower unitriangular*.

Proposition 2.2.5 ([25, Th. 2.10], [28, Lemma 3.2]). *The following conditions are equivalent for every* $M \in \mathcal{M}_n(\mathbb{SB})$:

   (i) *The column vectors* $M[\_, j]$ $(j \in \{1, \ldots, n\})$ *are independent;*
  (ii) *The row vectors* $M[i, \_]$ $(i \in \{1, \ldots, n\})$ *are independent;*
 (iii) Per $M = 1$;
 (iv) *M is congruent to some lower unitriangular matrix.*

Proof. (i) $\Rightarrow$ (iv). We use induction on n. Since the implication holds trivially for $n = 1$, we assume that $n > 1$ and the implication holds for $n - 1$. Assuming (i), it follows from Lemma 2.2.4 that M has a marker. By permuting the rows of M if needed, we may assume that the first row is a marker. By permuting columns if needed, we may assume that $M[1, \_] = (1, 0, \ldots, 0)$. Let $N = M[\bar{1}, \bar{1}]$. Then the column vectors $N[\_, j]$ $(j \in \{1, \ldots, n - 1\})$ are the column vectors $M[\_, j]$ $(j \in \{2, \ldots, n\})$ with the first coordinate removed. Since this first coordinate is always 0, it follows from Lemma 2.2.3(ii) that the column vectors of N are independent. By the induction hypothesis, N is congruent to some lower unitriangular matrix N', i.e. we can apply some sequence of row/column permutations to N to get N'. Now if we apply the same sequence of row/column permutations to the matrix M the first row of M remains unchanged, hence we obtain a lower unitriangular matrix as required.

   (iv) $\Rightarrow$ (iii). If $M = (m_{ij})$ is lower unitriangular, then the unique $\pi \in S_n$ such that $\prod_{i=1}^{n} m_{i,i\pi}$ is nonzero is the identity permutation. Hence Per $M = \prod_{i=1}^{n} m_{i,i} = 1$. Finally, we apply Proposition 2.2.2.

   (iii) $\Rightarrow$ (i). We use induction on n. Since the implication holds trivially for $n = 1$, we assume that $n > 1$ and the implication holds for $n - 1$. Note that, by Proposition 2.2.2, permuting rows or columns does not change the permanent. Clearly, the same happens with respect to the dependence of the column vectors.

   Suppose first that M has no marker. Since M cannot have a row consisting only of zeroes in view of Per $M = 1$, we have at least two nonzero entries in each row of M. Since Per $M = 1$, we may also assume, (independently) permuting rows and columns if necessary, that M has no zero entries on the main diagonal.

   We build a directed graph $\Gamma = (V, E)$ with vertex set $V = \{1, \ldots, n\}$ and edges $i \longrightarrow j$ whenever $m_{ij} \neq 0$. By our assumption on the main diagonal, we have a loop at each vertex i. Moreover, each vertex i must have outdegree at least two, each of the nonzero entries m_{ij} in the ith row producing an edge $i \longrightarrow j$. It follows that Γ must have a cycle

$$i_0 \longrightarrow i_1 \longrightarrow \ldots \longrightarrow i_k = i_0$$

of length $k \geq 2$ and so S_n contains two different permutations π, namely the identity and $(i_0\, i_1 \ldots i_{k-1})$, such that $m_{i,i\pi} \neq 0$ for every $i \in \{1, \ldots, n\}$. Hence Per $M = 2$, a contradiction.

Hence we may assume that $M = (m_{ij})$ has a marker. Permuting rows and columns if necessary, we may indeed assume that $M[1, \_] = (1, 0, \ldots, 0)$. Let $N = M[\bar{1}, \bar{1}]$. Then Proposition 2.2.1 yields Per $N = $ Per $M = 1$. By the induction hypothesis, the column vectors of N are independent. Suppose that

$$\lambda_1 M[\_, 1] + \ldots \lambda_n M[\_, n] \in \mathcal{G}^n$$

for some $\lambda_1, \ldots, \lambda_n \in \{0, 1\}$. Since $M[1, \_]$ is a marker, we get $\lambda_1 = 0$. Since the column vectors of N are independent, we get $\lambda_2 = \ldots = \lambda_n = 0$ as well. Thus the column vectors of M are independent.

(ii) $\Leftrightarrow$ (iii). Let M^t denote the transpose matrix of M. By Proposition 2.2.2(iii), we have

$$\text{Per } M = 1 \quad \Leftrightarrow \quad \text{Per } M^t = 1.$$

On the other hand, (ii) is equivalent to the column vectors $M^t[\_, j]$ ($j \in \{1, \ldots, n\}$) being independent. Now we use the equivalence (i) $\Leftrightarrow$ (iii). $\square$

A square matrix satisfying the above (equivalent) conditions is said to be *nonsingular*.

We consider now independence for any arbitrary nonempty subset of column vectors. Given (equipotent) nonempty $I, J \subseteq \{1, \ldots, n\}$, we say that I is a *witness* for J in M if $M[I, J]$ is nonsingular.

Proposition 2.2.6 ([25, Th. 3.11]). *The following conditions are equivalent for all* $M \in \mathcal{M}_{m \times n}(\mathbb{SB})$ *and* $J \subseteq \{1, \ldots, n\}$ *nonempty:*

(i) The column vectors $M[\_, j]$ ($j \in J$) are independent;
(ii) J has a witness in M.

Proof. (i) $\Rightarrow$ (ii). We use induction on $|J|$. Since the implication holds trivially for $|J| = 1$, we assume that $|J| > 1$ and the implication holds for smaller sets.

Applying Lemma 2.2.4 to the matrix $M' = M[\{1, \ldots, m\}, J]$, it follows that M' has a marker. In view of Proposition 2.2.2, (independently) permuting rows and columns does not compromise the existence of a witness, hence we may assume that $M'[1, \_] = (1, 0, \ldots, 0)$ and j_1 is the element of J corresponding to the first column of M'. Let $N = M'[\bar{1}, \bar{1}]$. Then the column vectors $N[\_, j]$ ($j \in \{1, \ldots, |J| - 1\}$) are the column vectors $M'[\_, j]$ ($j \in \{2, \ldots, |J|\}$) with the first coordinate removed. Since this first coordinate is always 0, it follows from

Lemma 2.2.3(ii) that the column vectors of N are independent. By the induction hypothesis, $J \setminus \{j_1\}$ has some witness I in N. Write $P = N[I, J \setminus \{j_1\}]$. Then $M[I \cup \{1\}, J]$ is of the form

$$\begin{pmatrix} 1 & 0 \\ \cdots & P \end{pmatrix}$$

and so Per $M[I \cup \{1\}, J] =$ Per $P = 1$. Hence $I \cup \{1\}$ is a witness for J in M.

(ii) $\Rightarrow$ (i). Assume that I is a witness for J in M. Let $N = M[I, J]$. By Proposition 2.2.5, the column vectors $N[\_, j]$ ($j \in \{1, \ldots, |J|\}$) are independent. Thus the vectors $M[\_, j]$ ($j \in J$) are independent by Lemma 2.2.3(i). $\square$

We can deduce a corollary on boolean matrices which will become useful in future chapters:

Corollary 2.2.7. *Let $M \in \mathcal{M}_{m \times n}(\mathbb{B})$ and let $M' \in \mathcal{M}_{(m+1) \times n}(\mathbb{B})$ be obtained by adding as an extra row the sum (in $\mathbb{B}$) of two rows of M. Then the following conditions are equivalent for every $J \subseteq \{1, \ldots, n\}$:*

(i) The column vectors $M[\_, j]$ ($j \in J$) are independent;
(ii) The column vectors $M'[\_, j]$ ($j \in J$) are independent.

Proof. (i) $\Rightarrow$ (ii). By Lemma 2.2.3(i).

(ii) $\Rightarrow$ (i). By Proposition 2.2.6, J has a witness I in M'. It is easy to see that if a marker u is the sum of some vectors in $\mathbb{B}^k$, then one of them is equal to u. Therefore, if the sum row occurs in $M'[I, J]$, we can always replace it by one of the summand rows and get a nonsingular matrix of the form $M[K, J]$. $\square$

We are now ready to introduce the notion of rank of a superboolean matrix:

Proposition 2.2.8 ([25, Th. 3.11]). *The following numbers coincide for a given $M \in \mathcal{M}_{m \times n}(\mathbb{SB})$:*

(i) The maximum number of independent column vectors in M;
(ii) The maximum number of independent row vectors in M;
(iii) The maximum size of a subset $J \subseteq \{1, \ldots, n\}$ having a witness in M;
(iv) The maximum size of a nonsingular submatrix of M.

Proof. Let $M\iota_1, \ldots, M\iota_4$ denote the integers defined by each of the conditions (i)–(iv) for M, respectively. The equality $M\iota_3 = M\iota_4$ follows from the definition of witness, and $M\iota_1 = M\iota_3$ follows from Proposition 2.2.6. Finally, $M\iota_2 = M'\iota_1 = M'\iota_4$. Since $M'\iota_4 = M\iota_4$ in view of Proposition 2.2.2(iii), we get $M\iota_2 = M\iota_4$. $\square$

The *rank* of a superboolean matrix M, denoted by rk M, is then the number given by any of the equivalent conditions of Proposition 2.2.8.

If M is a boolean matrix, we can still define rk M as its rank when viewed as a superboolean matrix. We note also that this notion of rank for boolean matrices does not coincide with the definition used by Berstel, Perrin and Reutenauer in [2, Section VI.3].

Chapter 3
Posets and Lattices

We study in this chapter boolean representations of posets, paying special attention to the case of lattices. Indeed, by considering $\vee$-generated lattices, we succeed in establishing a correspondence between boolean matrices and lattices which will be a cornerstone of the theory of boolean representations of simplicial complexes.

3.1 Basic Notions

For the various aspects of lattice theory, the reader is referred to [22, 23, 44].

All the posets in this book are finite, and we abbreviate $(P, \leq)$ to P if the partial order is implicit.

Let P be a poset and let $a, b \in P$. We say that a *covers* b if $a > b$ and there is no $c \in P$ satisfying $a > c > b$. We may describe P by means of its *Hasse diagram* Hasse P: this is a directed graph having vertex set P and edges $a \longrightarrow b$ whenever b covers a. If P is simple enough, it is common to draw Hasse P as an undirected graph, when the orientation of the edge $a \longrightarrow b$ is expressed by the fact that a is placed at a lower level than b in the picture.

For instance, if we order $\{1, \ldots, 10\}$ by (integer) division, we obtain the Hasse diagram

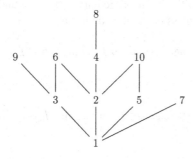

© Springer International Publishing Switzerland 2015
J. Rhodes, P.V. Silva, *Boolean Representations of Simplicial Complexes and Matroids*, Springer Monographs in Mathematics,
DOI 10.1007/978-3-319-15114-4_3

The *height* of P, denoted by ht P, is the maximum length of a chain in $\mathcal{P}$, i.e. the maximum number of edges in an (upward) path in Hasse P. For instance, the poset of the preceding example has height 3.

Given a poset P, we say that $X \subseteq P$ is a *down set* if

$$a \leq b \in X \quad \Rightarrow \quad a \in X$$

holds for all $a, b \in P$. Dually, X is an *up set* if

$$a \geq b \in X \quad \Rightarrow \quad a \in X$$

holds for all $a, b \in P$. The *principal* down set and up set generated by $a \in P$ are defined by

$$a\downarrow = \{x \in P \mid x \leq a\},$$
$$a\uparrow = \{x \in P \mid x \geq a\}.$$

A finite poset L is a *lattice* if there exist, for all $a, b \in L$, a *join* and a *meet* defined by

$$(a \vee_L b) = \min\{x \in L \mid x \geq a, b\},$$
$$(a \wedge_L b) = \max\{x \in L \mid x \leq a, b\}.$$

We drop the subscripts L whenever possible.

Since our lattices are finite, it follows that a lattice L has a minimum element B and a maximum element T. We refer to B as the *bottom* element and to T as the *top* element of L. We shall assume that $T \neq B$ in every lattice. A lattice is *trivial* if B and T are its unique elements.

Note also that finite lattices are always *complete* in the sense that

$$\vee_L S = \min\{x \in L \mid x \geq s \text{ for every } s \in S\},$$
$$\wedge_L S = \max\{x \in L \mid x \leq s \text{ for every } s \in S\},$$

are defined for every $S \subseteq L$. In particular, $\vee\emptyset = B$ and $\wedge\emptyset = T$. Another consequence of completeness is that the conditions

$$\text{there exists } a \vee b \text{ for all } a, b \in L \tag{3.1}$$

and

$$\text{there exists } a \wedge b \text{ for all } a, b \in L \tag{3.2}$$

are equivalent. Indeed, if (3.1) holds, then we have

$$(a \wedge b) = \vee\{x \in L \mid x \leq a, b\}. \tag{3.3}$$

We say that the meet (3.3) is *determined* by the join. Similarly, (3.2) implies (3.1), and

$$(a \vee b) = \wedge\{x \in L \mid x \geq a, b\}. \tag{3.4}$$

gives the determined join.

Given $S \subseteq L$, we say that S is a *sublattice* of L if

(SL1) $(a \vee_L b) \in S$ for all $a, b \in S$;
(SL2) $(a \wedge_L b) \in S$ for all $a, b \in S$;
(SL3) $B \in S$;
(SL4) $T \in S$.

Note that the first two conditions express the fact that S under the induced partial order is a lattice in its own right, with $(a \vee_S b) = (a \vee_L b)$ and $(a \wedge_S b) = (a \wedge_L b)$ for all $a, b \in S$.

If we only require conditions (SL1) and (SL3), we say that S is a $\vee$-*subsemilattice* of L. Since L is finite, this is equivalent to saying that $\vee_L X \in S$ for every $X \subseteq S$.

If we only require conditions (SL2) and (SL4), we say that S is a $\wedge$-*subsemilattice* of L. Since L is finite, this is equivalent to saying that $\wedge_L X \in S$ for every $X \subseteq S$.

If we require conditions (SL2), (SL3) and (SL4), we say that S is a *full* $\wedge$-*subsemilattice* of L, and dually for (SL1), (SL3), (SL4) and a *full* $\vee$-*subsemilattice*.

We denote the set of all $\vee$-subsemilattices (respectively $\wedge$-subsemilattices, full $\wedge$-subsemilattices) of L by $\mathrm{Sub}_\vee L$ (respectively $\mathrm{Sub}_\wedge L$, $\mathrm{FSub}_\wedge L$).

Let A be a finite set and consider 2^A ordered by inclusion. Then 2^A becomes a lattice with

$$(X \vee Y) = X \cup Y, \quad (X \wedge Y) = X \cap Y,$$

$T = A$ and $B = \emptyset$. Given $\mathcal{Z} \subseteq 2^A$, it is easy to see that

$$\hat{\mathcal{Z}} = \{\cap X \mid X \subseteq \mathcal{Z}\}$$

is the smallest $\wedge$-subsemilattice of 2^A containing $\mathcal{Z}$, i.e. the $\wedge$-subsemilattice of 2^A generated by $\mathcal{Z}$. Considering the determined join

$$(X \vee Y) = \cap\{Z \in \mathcal{Z} \mid X \cup Y \subseteq Z\},$$

$(\hat{\mathcal{Z}}, \subseteq)$ becomes indeed a lattice. However, $\hat{\mathcal{Z}}$ is not in general a sublattice of 2^A since the determined join $X \vee_{\hat{\mathcal{Z}}} Y$ needs not to coincide with $X \cup Y$, and the bottom element $\cap \mathcal{Z}$ needs not to be the emptyset.

We introduce also the notion of Rees quotient of a lattice, borrowed from semigroup theory (see [12]). Given a lattice L and a proper down set $I \subset L$, the *Rees quotient* L/I is the quotient of L by the equivalence relation $\sim_I$ defined on L by

$$x \sim_I y \quad \text{if} \quad x = y \text{ or } x, y \in I.$$

The elements of L/I are the equivalence class I and the singular equivalence classes $\{x\}$ $(x \in L \setminus I)$, which we identify with x.

Proposition 3.1.1. *Let L be a lattice and let I be a proper down set of L. Then L/I has a natural lattice structure.*

Proof. The partial ordering of L translates to L/I in the obvious way, with I as the bottom element. Clearly, L/I inherits a natural $\wedge$-semilattice structure, and then becomes a lattice with the determined join. $\square$

Note that the canonical projection $L \rightarrow L/I$ is a homomorphism of $\wedge$-semilattices, but not necessarily a lattice homomorphism.

Many important features and results of lattice theory can be unified under the concept of closure operator, which will play a major role in this monograph. Indeed, closure operators on the lattice $(2^V, \subseteq)$ will constitute an alternative to boolean matrices, as we shall see in Sect. 5.4.

Given a lattice L, we say that $\xi : L \rightarrow L$ is a *closure operator* if the following axioms hold for all $a, b \in L$:

(C1) $a \leq a\xi$;
(C2) $a \leq b \quad \Rightarrow \quad a\xi \leq b\xi$;
(C3) $a\xi = (a\xi)\xi$.

A subset $X \subseteq L$ is *closed* (with respect to ξ) if $X\xi = X$.

As we show in Sect. A.2 of the Appendix, closure operators are in some precise sense equivalent to other important lattice-theoretic concepts such as $\wedge$-subsemilattices or $\vee$-congruences.

We also introduce the following definitions, which we use throughout the monograph:

Let L, L' be finite lattices. Following the terminology of [44], we say that a mapping $\varphi : L \rightarrow L'$ is a:

- $\vee$-*morphism* if $(a \vee b)\varphi = (a\varphi \vee b\varphi)$ for all $a, b \in L$;
- $\wedge$-*morphism* if $(a \wedge b)\varphi = (a\varphi \wedge b\varphi)$ for all $a, b \in L$;
- $\vee$-*map* if $(\vee X)\varphi = \vee(X\varphi)$ for every $X \subseteq L$;
- $\wedge$-*map* if $(\wedge X)\varphi = \wedge(X\varphi)$ for every $X \subseteq L$.

It is easy to see, separating the cases of X being nonempty and empty, that φ is a $\vee$-map if and only if φ is a $\vee$-morphism and $B\varphi = B$. Similarly, φ is a $\wedge$-map if and only if φ is a $\wedge$-morphism and $T\varphi = T$.

An equivalence relation σ on a lattice L is said to be a

* $\vee$-*congruence* if $a\sigma b$ implies $(a \vee c)\sigma(b \vee c)$ for all $a, b, c \in L$;
* $\wedge$-*congruence* if $a\sigma b$ implies $(a \wedge c)\sigma(b \wedge c)$ for all $a, b, c \in L$.

Given a mapping $\varphi : X \to Y$, the *kernel* of φ is the equivalence relation on X defined by

$$\text{Ker}\,\varphi = \{(a, b) \in X \times X \mid a\varphi = b\varphi\}.$$

Next we import to the context of finite lattices a concept originated in semigroup theory with the purpose of decomposing $\vee$-maps. Such results will be applied later, namely in Sect. 5.5.

Let L be a finite lattice. An element $a \in L$ is said to be *strictly meet irreducible* (smi) if, for every $X \subseteq L$, $a = \wedge X$ implies $a \in X$. This is equivalent to saying that a is covered by exactly one element of L. Similarly, a is *strictly join irreducible* (sji) if, for every $X \subseteq L$, $a = \vee X$ implies $a \in X$. This is equivalent to saying that a covers exactly one element of L. We denote by smi(L) (respectively sji(L)) the set of all smi (respectively sji) elements of L.

An *atom* of a lattice is an element covering the bottom element B. Clearly, every atom is necessarily sji. We denote by at(L) the set of all atoms of L. Dually, a *coatom* is covered by the top element T, and is necessarily smi.

3.2 Representation of Posets

We introduce now the boolean representation of posets as defined by Izhakian and Rhodes in [30].

Given a poset P, let $M(P) = (m_{pq})$ denote the boolean $(P \times P)$-matrix defined by

$$m_{pq} = \begin{cases} 0 & \text{if } p \geq q \\ 1 & \text{otherwise} \end{cases} \tag{3.5}$$

Notice that this is not the standard boolean representation of posets, where $m_{pq} = 1$ if and only if $p \leq q$. In our definition, the pth row is the characteristic vector of the complement of the down set generated by p.

Furthermore, we shall view the boolean matrix $M(P)$ as a particular case of a superboolean matrix for the purpose of independence and rank. Since our concepts differ from the standard ones, we shall call them *c-independence* and *c-rank* to avoid any possible confusion. The letter c refers to the boolean operator $0^c = 1$, $1^c = 0$, since the roles of 0 and 1 in our representation are reversed with respect to the standard case (if we transpose the matrix).

In practice, we must of course agree on some fixed enumeration of the elements of P to make the correspondence with rows and columns. In most of our examples, this enumeration corresponds to the usual ordering of natural numbers.

More generally, if $P', P'' \subseteq P$, we denote by $M(P', P'')$ the $P' \times P''$ submatrix of $M(P)$.

In the next example, P is described by means of its Hasse diagram, and we consider the standard enumeration:

$$\text{Hasse } P: \qquad \begin{matrix} 3 & & 4 & & 5 \end{matrix} \tag{3.6}$$

$$M(P) = \begin{pmatrix} 0 & 1 & 1 & 1 & 1 \\ 1 & 0 & 1 & 1 & 1 \\ 0 & 0 & 0 & 1 & 1 \\ 0 & 0 & 1 & 0 & 1 \\ 1 & 1 & 1 & 1 & 0 \end{pmatrix}$$

Let P be a poset and write $M = M(P)$. We say that $p_1, \dots, p_k \in P$ are *c-independent* if the column vectors $M[\_, p_i]$ ($i \in \{1, \dots, k\}$) are independent (over $\mathbb{SB}$). The *c-rank* of P, denoted by c-rk P, is the maximum cardinality of a subset of c-independent elements. By Proposition 2.2.8, this is precisely the maximum size of a nonsingular submatrix of $M(P)$.

Applying Proposition 2.2.5, it is easy to check that the c-rank of the poset of (3.6) is 4. Indeed, the 4×4 submatrix $M[\overline{5}, \overline{5}]$ is nonsingular, but M itself is not since it has no marker.

3.3 ∨-Generating Subsets of Lattices

We say that a (finite) lattice L is ∨- *generated* by $A \subseteq L$ if $L = \{\vee X \mid X \subseteq A\}$. Note that we may always assume that $B \notin A$ since $B = \vee \emptyset$ anyway. Similarly, L is ∧-*generated* by A if $L = \{\wedge X \mid X \subseteq A\}$.

It is immediate that sji(L) (respectively smi(L)) constitutes the (unique) minimum ∨-generating set (respectively ∧-generating set) of L.

We denote by FLg the class of all ordered pairs (L, A), where L is a (finite) lattice ∨-generated by $A \subseteq L \setminus \{B\}$. We say that $(L, A), (L', A') \in$ FLg are *isomorphic* and we write $(L, A) \cong (L', A')$ if there exists a lattice isomorphism $L \to L'$ inducing a bijection $A \to A'$.

The following lemma is simple but important:

Lemma 3.3.1. *Let* $(L, A) \in$ FLg *and let* $x, y \in L$ *Then:*

(i) $x = \vee(x\downarrow \cap A)$;
(ii) $x \leq y$ *if and only if* $x\downarrow \cap A \subseteq y\downarrow \cap A$.

Proof. (i) Since $x \geq a$ for every $a \in x\downarrow \cap A$, then $x \geq \vee(x\downarrow \cap A)$. On the other hand, since L is $\vee$-generated by A, we have $x = \vee X$ for some $X \subseteq A$. Hence $X \subseteq x\downarrow \cap A$ and so $x = \vee X \leq \vee(x\downarrow \cap A)$.

(ii) The direct implication is obvious, and the converse follows from (i). □

Lemma 3.3.1 unveils already one of the advantages procured by $\vee$-generating subsets: the possibility of reducing the number of columns in boolean representations. Some other advantages will become evident later on.

Let $(L, A) \in$ FLg and let $M(L)$ be the boolean representation defined in Sect. 3.2. We define the boolean representation of (L, A) to be the $|L| \times |A|$ submatrix $M(L, A) = (m_{xa})$ of $M(L)$.

Hence

$$m_{xa} = \begin{cases} 0 & \text{if } x \geq a \\ 1 & \text{otherwise} \end{cases}$$

for all $x \in L$ and $a \in A$.

3.4 The Lattice of Flats of a Matrix

Adapting results from Izhakian and Rhodes [29], we associate in this section a lattice with a given boolean matrix.

Let $M = (a_{ij})$ be an $m \times n$ boolean matrix and let $C = \{1, \ldots, n\}$ denote the set of columns of M. For $i \in \{1, \ldots, m\}$, write

$$Z_i = \{j \in \{1, \ldots, n\} \mid a_{ij} = 0\} \in 2^C \tag{3.7}$$

and define

$$\mathcal{Z}(M) = \{Z_1, \ldots, Z_m\} \subseteq 2^C.$$

The *lattice of flats* of M is then the lattice Fl $M = \widehat{\mathcal{Z}(M)}$ defined in Sect. 3.1, having as elements the intersections of subsets of $\mathcal{Z}(M)$. This terminology is inspired by the applications to simplicial complexes and matroids (see Chap. 5): if M is a boolean matrix representing a matroid $\mathcal{H}$, then the flats of M are flats of $\mathcal{H}$ in the usual sense (Lemma 5.2.1). But the converse needs not to be true, since Fl H is far from being the unique lattice representing $\mathcal{H}$.

Now assume that M has no zero columns. This is equivalent to saying that $\emptyset \in$ Fl M. For $j \in \{1, \ldots, n\}$, define

$$Y_j = \cap \{Z_i \mid a_{ij} = 0\}$$

and let

$$\mathcal{Y}(M) = \{Y_1, \ldots, Y_n\} \subseteq \text{Fl } M.$$

Note that $Y_j = \cap\{Z_i \mid j \in Z_i\}$ and so $j \in Y_j$ for every j.

Proposition 3.4.1. *Let* $M = (a_{ij})$ *be an* $m \times n$ *boolean matrix without zero columns. Then* $(\text{Fl } M, \mathcal{Y}(M)) \in \text{FLg}$.

Proof. First note that Y_j can never be the bottom element $\emptyset$ since $j \in Y_j$. Hence it suffices to show that

$$Z_{i_1} \cap \ldots \cap Z_{i_k} = \vee \{Y_j \mid j \in Z_{i_1} \cap \ldots \cap Z_{i_k}\} \qquad (3.8)$$

holds for all $i_1, \ldots, i_k \in \{1, \ldots, m\}$.

Indeed, take $j \in Z_{i_1} \cap \ldots \cap Z_{i_k}$. On the one hand, we have $a_{i_1 j} = \ldots = a_{i_k j} = 0$ and so $Y_j \subseteq Z_{i_1} \cap \ldots \cap Z_{i_k}$. Thus $\vee\{Y_j \mid j \in Z_{i_1} \cap \ldots \cap Z_{i_k}\} \subseteq Z_{i_1} \cap \ldots \cap Z_{i_k}$.

On the other hand, since $j \in Y_j$ for every j, we get

$$Z_{i_1} \cap \ldots \cap Z_{i_k} \subseteq \cup\{Y_j \mid j \in Z_{i_1} \cap \ldots \cap Z_{i_k}\}$$
$$\subseteq \vee\{Y_j \mid j \in Z_{i_1} \cap \ldots \cap Z_{i_k}\}$$

and so (3.8) holds as required. $\square$

Hence $M \mapsto (\text{Fl } M, \mathcal{Y}(M))$ defines an operator from the set of boolean matrices without zero columns into FLg.

3.5 Matrices Versus Lattices

In this section, we relate the operators defined between lattices and boolean matrices in Sects. 3.3 and 3.4. We begin with the following remarks and we use all the notation from the preceding two sections:

Lemma 3.5.1. *Let* $(L, A) \in \text{FLg}$ *and let* $M = M(L, A) = (m_{xa})$. *Then, for all* $x, y \in L$:

(i) $Z_x = x {\downarrow} \cap A$;
(ii) $x = \vee Z_x$;
(iii) $x \leq y$ *if and only if* $Z_x \subseteq Z_y$.

Proof. (i) We have

$$Z_x = \{a \in A \mid m_{xa} = 0\} = \{a \in A \mid a \leq x\} = x{\downarrow} \cap A.$$

(ii) and (iii) follow from part (i) and Lemma 3.3.1. □

Next we show how we can recover $(L, A) \in$ FLg from the lattice of flats of its matrix representation.

Proposition 3.5.2. *Let* $(L, A) \in$ FLg *and let* $M = M(L, A)$. *Then:*

(i) $\mathcal{Y}(M) = \{Z_a \mid a \in A\}$;
(ii) $(\mathrm{Fl}\, M, \mathcal{Y}(M)) \cong (L, A)$.

Proof. (i) Write $M = (m_{xa})$. For every $a \in A$, we have

$$Y_a = \cap\{Z_x \mid m_{xa} = 0\} = \cap\{Z_x \mid a \leq x\} = Z_a$$

by Lemma 3.5.1(iii).

(ii) Let $\varphi : L \to \mathrm{Fl}\, M$ be defined by $x\varphi = Z_x$. By Lemma 3.5.1(iii), φ is a poset embedding. On the other hand, $a \leq (x \wedge y)$ if and only if $a \leq x$ and $a \leq y$, hence $Z_x \cap Z_y = Z_{x \wedge y}$ for all $x, y \in L$. This immediately generalizes to

$$Z_{x_1} \cap \ldots \cap Z_{x_k} = Z_{x_1 \wedge \ldots \wedge x_k} \tag{3.9}$$

for all $x_1, \ldots, x_k \in L$. Since $A = Z_T$, it follows that φ is surjective. Thus φ is an isomorphism of posets and therefore of lattices. □

We shall refer to $\mathrm{Fl}\, M(L, A)$ as the *lattice of flats* of $(L, A) \in$ FLg and we write $\mathrm{Fl}(L, A) = \mathrm{Fl}\, M(L, A)$. As we shall see later, $\mathrm{Fl}(L, A)$ can be in particular cases the lattice of flats of a matroid.

Corollary 3.5.3. *Let* $(L, A) \in$ FLg. *Then*

$$\mathrm{Fl}(L, A) = \{Z_x \mid x \in L\} = \{x{\downarrow} \cap A \mid x \in L\}.$$

Proof. The first equality follows from (3.9) and $A = Z_T$, the second from Lemma 3.5.1(i). □

In an effort to characterize the boolean matrices arising as representations of some $(L, A) \in$ FLg, we now consider five properties for a boolean matrix M.

(M1) The rows of M are all distinct;
(M2) The columns of M are all distinct;
(M3) M contains a row with all entries equal to 0;
(M4) M contains a row with all entries equal to 1;
(M5) The set of row vectors of M is closed under addition in $\mathbb{B}^{|A|}$.

Let $\mathcal{M}$ denote the set of all boolean matrices satisfying properties (M1)–(M5).

Given a boolean matrix M without zero columns, we write

$$M^{\mu} = M(\text{Fl}\, M, \mathcal{Y}(M)).$$

In general, M^{μ} needs not to be congruent to M, as it will become apparent after Proposition 3.5.5. But we obtain better results for matrices in $\mathcal{M}$:

Proposition 3.5.4. *Let $M \in \mathcal{M}$. Then $M^{\mu} \cong M$.*

Proof. Assume that $M = (a_{ij})$ is an $m \times n$ matrix in $\mathcal{M}$. By definition, the elements of Fl M are of the form $\cap W$ for $W \subseteq \{Z_1, \ldots, Z_m\}$. Since M satisfies (M5), $\{Z_1, \ldots, Z_m\}$ is closed under intersection. In view of (M3), we have also $\cap\emptyset \in \{Z_1, \ldots, Z_m\}$, hence Fl $M = \{Z_1, \ldots, Z_m\}$. By (M1), these elements are all distinct. Note that, by (M4), M has no zero columns and so $(\text{Fl}\, M, \mathcal{Y}(M)) \in \text{FLg}$ by Proposition 3.4.1.

Therefore $M^{\mu} = (a'_{Z_i Y_j})$ is also a boolean matrix with m rows. To complete the proof, it suffices to show that $a'_{Z_i Y_j} = a_{ij}$ for all $i \in \{1, \ldots, m\}$ and $j \in \{1, \ldots, n\}$. In view of (M2), M^{μ} is then an $m \times n$ matrix in $\mathcal{M}$ and we shall be done.

Indeed, $a'_{Z_i Y_j} = 0$ if and only if $Y_j \subseteq Z_i$. Since $j \in Y_j$, this implies $j \in Z_i$. Conversely, $j \in Z_i$ implies $Y_j \subseteq Z_i$ and so

$$a'_{Z_i Y_j} = 0 \Leftrightarrow Y_j \subseteq Z_i \Leftrightarrow j \in Z_i \Leftrightarrow a_{ij} = 0.$$

Therefore $a'_{Z_i Y_j} = a_{ij}$ and so $M^{\mu} \cong M$. $\square$

We can now prove the following:

Proposition 3.5.5. *The following conditions are equivalent for a boolean matrix M:*

(i) $M = M(L, A)$ for some $(L, A) \in \text{FLg}$;
(ii) $M \in \mathcal{M}$.

Proof. (i) $\Rightarrow$ (ii). Write $M = (m_{xa})$. Property (M1) follows from Lemma 3.5.1(iii).

Let $a, b \in A$ be distinct. We may assume that $a \not\leq b$. Then $m_{ba} = 1 \neq 0 = m_{bb}$. Hence the columns corresponding to a and b are different and (M2) holds.

Property (M3) follows from $M[T, \_]$ being the zero vector. Since $B \notin A$, we get $M[B, \_] = (1, \ldots, 1)$ and so (M4) holds.

To prove (M5), let $x, y \in L$. It suffices to show that $m_{x \wedge y, a} = m_{x, a} + m_{y, a}$ holds in $\mathbb{B}$ for every $a \in A$. This follows from the equivalence

$$m_{x \wedge y, a} = 0 \Leftrightarrow a \leq (x \wedge y) \Leftrightarrow (a \leq x \text{ and } a \leq y)$$
$$\Leftrightarrow (m_{x, a} = 0 \text{ and } m_{y, a} = 0) \Leftrightarrow m_{x, a} + m_{y, a} = 0.$$

Therefore $M \in \mathcal{M}$.

(ii) $\Rightarrow$ (i). By Propositions 3.4.1 and 3.5.4. $\square$

Now it is easy to establish a correspondence between the set FLg/ $\cong$ of isomorphism classes of FLg and the set $\mathcal{M}/\cong$ of congruence classes of $\mathcal{M}$:

Corollary 3.5.6. *The mappings*

$$\mathcal{M} \to \text{FLg} \qquad and \qquad \text{FLg} \to \mathcal{M}$$
$$M \mapsto (\text{Fl } M, \mathcal{Y}(M)) \qquad\qquad (L, A) \mapsto M(L, A)$$

induce mutually inverse bijections between $\mathcal{M}/\cong$ and FLg/ $\cong$.

Proof. It follows easily from the definitions that the above operators induce mappings between $\mathcal{M}/\cong$ and FLg/ $\cong$. These mappings are mutually inverse by Propositions 3.5.2(ii) and 3.5.4. $\square$

Example 3.5.7. Let M be the matrix

$$\begin{pmatrix} 1 & 0 & 1 & 0 & 1 \\ 1 & 0 & 0 & 1 & 1 \\ 1 & 1 & 0 & 0 & 0 \end{pmatrix}$$

Omitting brackets and commas, and identifying the elements $Y_1, \ldots, Y_5$ of $\mathcal{Y}(M)$, the lattice of flats *Fl* M can be represented through its Hasse diagram:

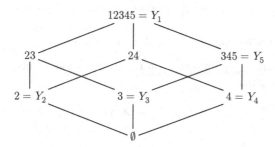

Finally, M^μ is the matrix

$$\begin{pmatrix} 1 & 0 & 1 & 0 & 1 \\ 1 & 0 & 0 & 1 & 1 \\ 1 & 1 & 0 & 0 & 0 \\ 1 & 0 & 1 & 1 & 1 \\ 1 & 1 & 0 & 1 & 1 \\ 1 & 1 & 1 & 0 & 1 \\ 0 & 0 & 0 & 0 & 0 \\ 1 & 1 & 1 & 1 & 1 \end{pmatrix}$$

The above example illustrates a simple remark: if all the columns of M are distinct and nonzero, if all its rows are distinct, then M^μ can be obtained from M by adding a zero row and any new rows obtained as sums of rows of M in $\mathcal{B}^{|E|}$.

3.6 c-Independence and c-Rank

Let L be a lattice and $M = M(L)$. As defined in Sect. 3.2, the elements $x_1, \ldots, x_k \in L$ are c-independent if the column vectors $M[\_, x_1], \ldots, M[\_, x_k]$ are independent (over $\mathbb{SB}$). Note that, if $(L, A) \in$ FLg and $x_1, \ldots, x_k \in A$, this is equivalent to saying that the column vectors $M'[\_, x_1], \ldots, M'[\_, x_k]$ of $M' = M(L, A)$ are independent (over $\mathbb{SB}$).

We can use the lattice of flats $\mathrm{Fl}(L, A)$ to define an operator on the lattice $(2^A, \subseteq)$: given $X \subseteq A$, let

$$\mathrm{Cl}_L X = \cap \{Z \in \mathrm{Fl}(L, A) \mid X \subseteq Z\}.$$

Lemma 3.6.1. *Let $(L, A) \in$ FLg. Then Cl_L is a closure operator on $(2^A, \subseteq)$.*

Proof. By construction, $\mathrm{Fl}(L, A)$ is a $\cap$-subsemilattice of $(2^A, \subseteq)$, and

$$\mathrm{Cl}_L = (\mathrm{Fl}(L, A))\Phi'$$

is a closure operator by Proposition A.2.4 in the Appendix. □

Recall the notation Z_i introduced in (3.7) for a boolean matrix M. If $M = M(L, A)$, we claim that

$$\mathrm{Cl}_L X = Z_{\vee_L X} = A \cap (\vee_L X)\!\downarrow \tag{3.10}$$

holds for every $X \subseteq A$.

Indeed, we have $X \subseteq Z_{\vee_L X} \in \mathrm{Fl}(L, A)$, and the equivalence

$$X \subseteq Z_y \quad \Leftrightarrow \quad \forall x \in X \; x \leq y \quad \Leftrightarrow \quad \vee_L X \leq y \quad \Leftrightarrow \quad Z_{\vee_L X} \subseteq Z_y$$

follows from Lemma 3.5.1, hence $\mathrm{Cl}_L X = Z_{\vee_L X}$. Lemma 3.5.1 also yields $Z_{\vee_L X} = A \cap (\vee_L X)\!\downarrow$, hence (3.10) holds.

The *successive differences* of a chain $Y_0 \supset \ldots \supset Y_k$ in 2^A are the subsets $Y_0 \setminus Y_1, \ldots, Y_{k-1} \setminus Y_k$. If $Y_0 = A$ and $Y_k = \emptyset$, they constitute an (ordered) partition of A. We say that $X = \{x_1, \ldots, x_k\} \subseteq A$ is a *transversal* of the successive differences for the above chain if each $Y_{i-1} \setminus Y_i$ contains exactly one element of X. A subset of a transversal is a *partial transversal* .

By adapting the proofs of [29, Lemmas 3.4 and 3.5], we can prove the following:

Proposition 3.6.2. *Let* $(L, A) \in \mathrm{FLg}$ *and* $X \subseteq A$. *Then the following conditions are equivalent:*

(i) X *is c-independent as a subset of* L;
(ii) X *admits an enumeration* $x_1, \ldots, x_k$ *such that*

$$(x_1 \vee \ldots \vee x_k) > (x_2 \vee \ldots \vee x_k) > \ldots > (x_{k-1} \vee x_k) > x_k; \qquad (3.11)$$

(iii) X *admits an enumeration* $x_1, \ldots, x_k$ *such that*

$$\mathrm{Cl}_L(x_1, \ldots, x_k) \supset \mathrm{Cl}_L(x_2, \ldots, x_k) \supset \ldots \supset \mathrm{Cl}_L(x_k);$$

(iv) X *admits an enumeration* $x_1, \ldots, x_k$ *such that*

$$x_i \notin \mathrm{Cl}_L(x_{i+1}, \ldots, x_k) \qquad (i = 1, \ldots, k-1);$$

(v) X *is a transversal of the successive differences for some chain of* $\mathrm{Fl}(L, A)$;
(vi) X *is a partial transversal of the partition of successive differences for some maximal chain of* $\mathrm{Fl}(L, A)$.

Proof. (i) $\Rightarrow$ (ii). If X is c-independent, then X admits an enumeration $x_1, \ldots, x_k$ such that $M(L)$ admits a lower unitriangular submatrix M', with the columns labelled by $x_1, \ldots, x_k$ and the rows labelled, say, by $y_1, \ldots, y_k$. Let $i \in \{1, \ldots, k-1\}$. We have $x_i \not\leq y_i$ and $x_{i+1}, \ldots, x_k \leq y_i$ since M' is lower unitriangular. Hence $(x_{i+1} \vee \ldots \vee x_k) \leq y_i$ but $(x_i \vee \ldots \vee x_k) \not\leq y_i$ in view of $x_i \not\leq y_i$. Thus $(x_i \vee \ldots \vee x_k) > (x_{i+1} \vee \ldots \vee x_k)$ and (3.11) holds.

(ii) $\Rightarrow$ (iii). By Lemma 3.5.1(iii) and (3.10).

(iii) $\Rightarrow$ (iv). If $x_i \in \mathrm{Cl}_L(x_{i+1}, \ldots, x_k)$, then $\mathrm{Cl}_L(x_i, \ldots, x_k) = \mathrm{Cl}_L(x_{i+1}, \ldots, x_k)$.

(iv) $\Rightarrow$ (ii). Clearly, $(x_i \vee \ldots \vee x_k) \geq (x_{i+1} \vee \ldots \vee x_k)$, and equality would imply $\mathrm{Cl}_L(x_i, \ldots, x_k) = \mathrm{Cl}_L(x_{i+1}, \ldots, x_k)$ by (3.10).

(ii) $\Rightarrow$ (i). If (3.11) holds, we build a lower unitriangular submatrix of $M(L)$ by taking rows labelled by $y_1, \ldots, y_k \in L$, where $y_i = (x_{i+1} \vee \ldots \vee x_k)$ $(i = 1, \ldots, k)$.

(v) $\Rightarrow$ (iv). We may assume that there exists a chain

$$Y_0 \supset \ldots \supset Y_k$$

in $\mathrm{Fl}(L, A)$ and an enumeration $x_1, \ldots, x_k$ of the elements of X such that $x_i \in Y_{i-1} \setminus Y_i$ for $i = 1, \ldots, k$. Since $\mathrm{Cl}_L(x_{i+1}, \ldots, x_k) \subseteq Y_i$, it follows that $x_i \notin \mathrm{Cl}_L(x_{i+1}, \ldots, x_k)$ for $i = 1, \ldots, k-1$.

(iii) $\Rightarrow$ (v). It is easy to see that X is a transversal of the successive differences for the chain

$$\mathrm{Cl}_L(x_1, \ldots, x_k) \supset \mathrm{Cl}_L(x_2, \ldots, x_k) \supset \ldots \supset \mathrm{Cl}_L(x_{k-1}, x_k) \supset \mathrm{Cl}_L(x_k) \supset \emptyset$$

in $\mathrm{Fl}(L, A)$.

(v) $\Leftrightarrow$ (vi). Immediate. $\square$

It is easy to characterize c-independence for small numbers of elements:

Corollary 3.6.3. *Let* $(L, A) \in$ FLg *and let* $X \subseteq A$ *with* $|X| \leq 2$. *Then* X *is c-independent as a subset of* L.

Proof. The case $|X| \leq 1$ is immediate in view of $B \notin A$, hence we may assume that $X = \{x_1, x_2\}$ and $x_1 \not\leq x_2$. Then $(x_1 \vee x_2) > x_2$ and so X is c-independent by Proposition 3.6.2. $\square$

We end this section by discussing the c-rank. The second equality is due to Izhakian and Rhodes [29, Theorem 3.6].

Proposition 3.6.4. *Let* $(L, A) \in$ FLg. *Then* rk $M(L, A) =$ c-rk $L =$ ht L.

Proof. Since $M(L, A)$ is a submatrix of $M(L)$, we have rk $M(L, A) \leq$ c-rk L.

Suppose that ht $L = k$. Then there is a chain $y_0 > y_1 > \ldots > y_k$ in L. Since A $\vee$-generates L, for each $i \in \{1, \ldots, k\}$ there exists some $a_i \in A$ such that $a_i \leq y_{i-1}$ but $a_i \not\leq y_i$. It follows that

$$(a_{i+1} \vee \ldots \vee a_k) \leq y_i \not\geq (a_i \vee \ldots \vee a_k)$$

and so

$$(a_1 \vee \ldots \vee a_k) > (a_2 \vee \ldots \vee a_k) > \ldots > (a_{k-1} \vee x_k) > a_k.$$

By Proposition 3.6.2, $\{a_1, \ldots, a_k\}$ is c-independent and so rk $M(L, A) \geq$ ht L.

Now c-rk $L =$ rk $M(L) =$ rk $M(L, L \setminus \{B\})$ because the omitted column corresponding to B contains only zeros and is therefore irrelevant to the computation of the c-rank. By Proposition 3.6.2, every c-independent subset X of $L \setminus \{B\}$ with k elements produces a chain in L with length $k - 1$, which can be extended to a chain of length k by adjoining the bottom element B. Hence c-rk $L \leq$ ht $L \leq$ rk $M(L, A) \leq$ c-rk L and we are done. $\square$

In Sect. A.4 of the Appendix, we use the results in this section to produce results on c-independence and c-rank for posets (Propositions A.4.3 and A.4.4) using the concept of lattice completion.

Chapter 4
Simplicial Complexes

Simplicial complexes can be approached in the most abstract way, as combinatorial objects, and under this perspective they are often called hereditary collections. Matroids constitute a very important particular case.

They can be also approached under a geometric perspective (say, as objects in some Euclidean space), and this leads to the concept of geometric simplicial complex.

The two definitions turn out to be equivalent in some sense. In this monograph, we shall favor the combinatorial perspective on most occasions, but we shall also use the geometric perspective for some results. We note that all our simplicial complexes are assumed to be finite.

In this chapter, we focus on the combinatorial perspective and discuss the fundamental concept of *flat*, which generalizes the well-known concept of matroid theory and is bound to play a major role in the upcoming theory of boolean representations.

The geometric perspective is discussed in some detail in Sect. A.5 of the Appendix.

4.1 The Combinatorial Perspective

Let V be a finite nonempty set and let $H \subseteq 2^V$. We say that $\mathcal{H} = (V, H)$ is an (abstract) *simplicial complex* (or *hereditary collection*) if H is nonempty and closed under taking subsets. Two simplicial complexes $\mathcal{H} = (V, H)$ and $\mathcal{H}' = (V', H')$ are *isomorphic* if there exists some bijection $\varphi : V \to V'$ such that

$$X \in H \Leftrightarrow X\varphi \in H'$$

holds for every $X \subseteq V$. We write then $\mathcal{H} \cong \mathcal{H}'$.

© Springer International Publishing Switzerland 2015
J. Rhodes, P.V. Silva, *Boolean Representations of Simplicial Complexes and Matroids*, Springer Monographs in Mathematics,
DOI 10.1007/978-3-319-15114-4_4

A nonempty element $I \in H$ is called a *simplex* or a *face*. Its *dimension* is $\dim I = |I| - 1$. A maximal face of $\mathcal{H}$ (under inclusion) is called a *facet*. We denote by fct $\mathcal{H}$ the set of all facets of $\mathcal{H}$.

We use the convention $\dim \emptyset = -1$. Then the dimension of $\mathcal{H}$ is defined by

$$\dim \mathcal{H} = \max\{\dim X \mid X \in H\}.$$

For every $k \geq 0$, we write

$$P_k(V) = \{X \subseteq V \mid |X| = k\} \quad \text{and} \quad P_{\leq k}(V) = \{X \subseteq V \mid |X| \leq k\}.$$

We may on occasions identify $P_1(V)$ with V. We shall refer to their elements as *points*. We adopt also on occasions the terminology *n-set* (respectively *n-subset*) to refer to a set (respectively subset) with n elements. To simplify notation, we shall often represent an n-set $\{x_1, x_2, \ldots, x_n\}$ as $x_1 x_2 \ldots x_n$.

A simplicial complex $\mathcal{H} = (V, H)$ is:

- *Pure* if all its facets have the same dimension;
- *Trim* if $P_1(V) \subseteq H$;
- *Simple* if $P_2(V) \subseteq H$;
- *Paving* if $P_d(V) \subseteq H$ for $d = \dim \mathcal{H}$;
- *Uniform* if $H = P_{\leq d+1}(V)$.

A simplicial complex $\mathcal{H} = (V, H)$ satisfies the *point replacement property* if:

(PR)   for all $I, \{p\} \in H \setminus \{\emptyset\}$, there exists some $i \in I$ such that $(I \setminus \{i\}) \cup \{p\} \in H$.

4.1.1 Matroids

A simplicial complex $\mathcal{H} = (V, H)$ satisfies the *exchange property* if:

(EP)   For all $I, J \in H$ with $|I| = |J| + 1$, there exists some $i \in I \setminus J$ such that $J \cup \{i\} \in H$.

A simplicial complex satisfying the exchange property is called a *matroid*. Note that every matroid is necessarily pure. Moreover, (EP) implies the more general condition

(EP')   For all $I, J \in H$ with $|I| > |J|$, there exists some $i \in I \setminus J$ such that $J \cup \{i\} \in H$.

Indeed, it suffices to apply (EP) to J and I' for some $I' \subseteq I$ such that $|I'| = |J| + 1$.

There are many other equivalent definitions of matroid. For details, the reader is referred to [39, 40, 54]. The concept of *circuit* is very important. A subset $C \subseteq V$ is said to be a circuit of $\mathcal{H} = (V, H)$ if $C \notin H$ but all proper subsets of C are in H.

4.1.2 Graphs

We can view (finite undirected) graphs as trim simplicial complexes of dimension ≤ 1. If $\mathcal{H} = (V, H)$ is such a complex, we view V (identified often with $P_1(V)$) as the set of *vertices* and $E = P_2(V) \cap H$ as the set of *edges*. However, we shall use the classical representation (V, E) to denote a graph. Note that this definition excludes the existence of loops or multiple edges, so we are meaning graphs in the strictest sense of the word.

Graphs will be present throughout the whole monograph, and we collect here some of the concepts and terminology to be needed in future sections. We assume some familiarity with the most basic concepts such as *subgraph, isomorphism, path, cycle* and *degree*.

Let $\Gamma = (V, E)$ be a (finite undirected) graph. The *girth* of G, denoted by gth Γ, is the length of the shortest cycle in Γ (assumed to be ∞ if Γ is acyclic). Note that gth $\Gamma \geq 3$ since there are no loops nor multiple edges. A graph of girth >3 is said to be *triangle-free*.

The graph Γ is *connected* if all distinct $v, w \in V$ can be connected through some path of the form

$$v = v_0 - v_1 - \ldots - v_n = w,$$

where $v_{i-1} v_i \in E$ for every $i \in \{1, \ldots, n\}$. Such a path is said to have *length n*.

We shall use the expression *connected component* of Γ to denote both a maximal connected subgraph of Γ and its set of vertices. We say that a connected component is nontrivial if it has more than one vertex (equivalently, if it has an edge).

If Γ is a connected graph, we can define a metric d on V by taking $d(v, w)$ to be the length of the shortest path connecting v and w. The *diameter* of Γ is defined as

$$\text{diam } \Gamma = \max\{d(v, w) \mid v, w \in V\}.$$

If Γ is not connected, we define diam $\Gamma = \infty$.

We say that $X \subseteq V$ is a

- *Clique* if $pq \in E$ for all distinct $p, q \in X$;
- *Anticlique* if $pq \notin E$ for all distinct $p, q \in X$.

Anticliques are more usually called independent sets, but we want to avoid overloading the word "independent". An (anti)clique with n elements is said to be an n-(anti)clique. A 1-(anti)clique is said to be trivial.

Given a vertex $v \in V$, the *neighborhood* of v is defined as the set nbh(v) of all vertices adjacent to v in Γ. The *closed neighborhood* of v is defined as $\overline{\text{nbh}}(v) = \text{nbh}(v) \cup \{v\}$.

We denote by maxdeg Γ the maximum degree reached by a vertex in Γ, i.e. maxdeg $\Gamma = \max\{|\text{nbh}(v)| \mid v \in V\}$.

The *complement* $\Gamma^c = (V, E^c)$ of Γ is the graph defined by $E^c = P_2(V) \setminus E$.

We introduce next the concepts of *superclique* and *superanticlique* for a graph $\Gamma = (V, E)$. We say that a nontrivial clique $C \subseteq V$ is a *superclique* if

$$\overline{\text{nbh}}(a) \cap \overline{\text{nbh}}(b) = C$$

holds for all $a, b \in C$ distinct. In particular, every superclique is a maximal clique. Dually, a nontrivial anticlique $A \subseteq V$ is a *superanticlique* if

$$\text{nbh}(a) \cup \text{nbh}(b) = V \setminus A$$

holds for all $a, b \in A$ distinct. In particular, every superanticlique is a maximal anticlique.

Proposition 4.1.1. *Let $\Gamma = (V, E)$ be a graph and let $X \subseteq V$. Then the following conditions are equivalent:*

(i) X is a superanticlique of Γ;
(ii) X is a superclique of Γ^c.

Proof. Indeed, it is immediate that cliques on Γ correspond to anticliques in Γ^c and vice-versa. For the rest, we compute the complements. Clearly, $\text{nbh}_\Gamma(p) = V \setminus \overline{\text{nbh}}_{\Gamma^c}(p)$, and $\overline{\text{nbh}}_{\Gamma^c}(p) \cap \overline{\text{nbh}}_{\Gamma^c}(q) = X$ if and only if $\text{nbh}_\Gamma(p) \cup \text{nbh}_\Gamma(q) = V \setminus X$. $\square$

Given graphs Γ and Γ', we denote by $\Gamma \sqcup \Gamma'$ their *disjoint union*.

Given $n \geq 1$, we denote by K_n the *complete* graph on n vertices, i.e. we take $|V| = n$ and $E = P_2(V)$. Given $m, n \geq 1$, we denote by $K_{m,n}$ the *complete bipartite graph* on $m + n$ vertices, i.e. we consider a partition $V = V_1 \cup V_2$ such that $|V_1| = m$, $|V_2| = n$ and

$$E = \{v_1 v_2 \mid v_1 \in V_1, v_2 \in V_2\}.$$

4.2 Flats

Let $\mathcal{H} = (V, H)$ be a simplicial complex. We say that $X \subseteq V$ is a *flat* if

$$\forall I \in H \cap 2^X \quad \forall p \in V \setminus X \qquad I \cup \{p\} \in H.$$

The set of all flats of $\mathcal{H}$ is denoted by $\text{Fl } \mathcal{H}$.

An alternative characterization is provided through the notion of circuit:

Proposition 4.2.1. *Let $\mathcal{H} = (V, H)$ be a simplicial complex and let $X \subseteq V$. Then the following conditions are equivalent:*

(i) X is a flat;
(ii) If $p \in C \subseteq X \cup \{p\}$ for some circuit C, then $p \in X$.

Proof. (i) $\Rightarrow$ (ii). Suppose that there exist a circuit C and $p \in C \subseteq X \cup \{p\}$ such that $p \notin X$. Then $C = I \cup \{p\}$ for some $I \subseteq X$. It follows that $I \in H \cap 2^X$ and $p \in V \setminus X$, however $I \cup \{p\} \notin H$. Therefore X is not a flat.

(ii) $\Rightarrow$ (i). Suppose that X is not a flat. Then there exist $I \in H \cap 2^X$ and $p \in V \setminus X$ such that $I \cup \{p\} \notin H$. Let $I_0 \subseteq I$ be minimal for the property $I_0 \cup \{p\} \notin H$. Since $I_0 \in H$ due to $I_0 \subseteq I \in H$, it follows that $I_0 \cup \{p\}$ is a circuit by minimality of I_0. Thus condition (ii) fails for $C = I_0 \cup \{p\}$ and we are done. $\square$

Note that condition (ii) is one of the standard characterization of flats for matroids.

The following result summarizes some straightforward properties of Fl $\mathcal{H}$.

Proposition 4.2.2. *Let $\mathcal{H} = (V, H)$ be a simplicial complex.*

(i) $V \in$ Fl $\mathcal{H}$.
(ii) *If $Y \subseteq$ Fl$\mathcal{H}$, then $\cap Y \in$ Fl$\mathcal{H}$.*
(iii) *If $P_{\leq k}(V) \subseteq H$ for $k \geq 1$, then $P_{\leq k-1}(V) \subseteq$ Fl$\mathcal{H}$.*
(iv) *If $\mathcal{H}$ is simple, then the points of V are flats.*

Proof. (i) Trivial.
(ii) In view of part (i), it suffices to show that $X_1, X_2 \in$ Fl $\mathcal{H}$ implies $X_1 \cap X_2 \in$ Fl $\mathcal{H}$.
    Let $I \in H \cap 2^{X_1 \cap X_2}$ and $p \in V \setminus (X_1 \cap X_2)$. We may assume that $p \notin X_1$. Since $I \in H \cap 2^{X_1}$ and X_1 is a flat, it follows that $I \cup \{p\} \in H$. Thus $X_1 \cap X_2 \in$ Fl $\mathcal{H}$.
(iii) Immediate from the definition.
(iv) By part (iii). $\square$

By part (ii) of the preceding proposition, Fl $\mathcal{H}$ is a $\cap$-subsemilattice of $(2^V, \subseteq)$, and so we may define a closure operator on 2^V by

$$\text{Cl } X = \cap\{Z \in \text{Fl } \mathcal{H} \mid X \subseteq Z\}$$

(i.e. Cl $= (\text{Fl } \mathcal{H})\Phi'$, see Proposition A.2.4 in the Appendix). We shall also use the notation $\overline{X} = \text{Cl } X$. Note that $X \subseteq V$ is closed if and only if Cl $X = X$ if and only if X is a flat. We shall also refer to flats as *closed subsets*.

We can also make the following remark:

Proposition 4.2.3. *Let $\mathcal{H} = (V, H)$ be a simplicial complex and let $X \subseteq V$ be a facet. Then Cl $X = V$.*

Proof. Suppose that $p \in V \setminus \text{Cl } X$. Since $X \in H \cap 2^{\text{Cl } X}$ and Cl X is a flat, we get $X \cup \{p\} \in H$, contradicting X being a facet. Thus Cl $X = V$. $\square$

We can obtain Fl $\mathcal{H}$ as closures of faces.

Proposition 4.2.4. *Let $\mathcal{H} = (V, H)$ be a simplicial complex. Then Fl $\mathcal{H} = \{\text{Cl } X \mid X \in H\}$.*

Proof. Let $F \in \text{Fl } \mathcal{H}$ and let $I \in H \cap 2^F$ be maximal. By maximality of I, we have $I \cup \{p\} \notin H$ for every $p \in F \setminus I$. Thus every flat containing I must contain F and so $F = \text{Cl } I$. By Proposition 4.2.3, we get $I \notin \text{fct } \mathcal{H}$ and the direct inclusion. The opposite inclusion is trivial. $\square$

The matroid case allows for a more constructive computation of the closure. Given a simplicial complex $\mathcal{H} = (V, H)$ and let $X \subseteq V$, we define

$$X\delta = X \cup \{p \in V \setminus X \mid I \cup \{p\} \notin H \text{ for some } I \in H \cap 2^X\}.$$

In the matroid case, δ defines the closure. In the general case, we have to iterate.

Proposition 4.2.5. *Let $\mathcal{H} = (V, H)$ be a simplicial complex and let $X \subseteq V$. Then:*

(i) $\text{Cl } X = \cup_{n \geq 0} X\delta^n$;
(ii) If $\mathcal{H}$ is a matroid, then $\text{Cl } X = X\delta$.

Proof. (i) It is easy to see that

$$X\delta \subseteq \text{Cl } X \tag{4.1}$$

holds for every $X \subseteq V$:

Let $p \in X\delta \setminus X$. Then $I \cup \{p\} \notin H$ for some $I \in H \cap 2^X \subseteq H \cap 2^{\text{Cl } X}$. If $p \notin \text{Cl } X$, this contradicts $\text{Cl } X$ being closed, hence $p \in \text{Cl } X$ and so $X\delta \subseteq \text{Cl } X$.

We show that $X\delta^n \subseteq \text{Cl } X$ holds for every $n \geq 0$ by induction. The case $n = 0$ being trivial, assume that $X\delta^k \subseteq \text{Cl } X$. Then $X\delta^{k+1} \subseteq \text{Cl}(X\delta^k)$ by (4.1) and so $X\delta^{k+1} \subseteq \text{Cl } X$ since $X\delta^k \subseteq \text{Cl } X$. Therefore $\cup_{n \geq 0} X\delta^n \subseteq \text{Cl } X$.

It remains to be proved that $Y = \cup_{n \geq 0} X\delta^n$ is closed. Let $p \in V \setminus Y$ and let $I \in H \cap 2^Y$. Since $X \subseteq X\delta \subseteq X\delta^2 \subseteq \ldots$ and I is finite, we have $I \in H \cap 2^{X\delta^m}$ for some $m \geq 1$.

Suppose that $I \cup \{p\} \notin H$. Since $p \in V \setminus X\delta^m$, then $p \in X\delta^{m+1}$, contradicting $p \in V \setminus Y$. Thus $I \cup \{p\} \in H$ and so Y is closed. Therefore $\text{Cl } X = Y$.

(ii) For every $Z \subseteq V$, let $H(Z)$ denote the set of all faces of $\mathcal{H}$ contained in Z with maximum possible dimension. We start by proving that

$$X\delta = X \cup \{p \in V \setminus X \mid J \cup \{p\} \notin H \text{ for every } J \in H(X)\}. \tag{4.2}$$

Indeed, let $p \in X\delta \setminus X$. Then $I \cup \{p\} \notin H$ for some $I \in H \cap 2^X$. Since we may replace I by any $I' \in H \cap 2^X$ containing I, We may assume that I is maximal with respect to $I \in H \cap 2^X$ and $I \cup \{p\} \notin H$.

Take $J \in H(X)$ and suppose that $J \cup \{p\} \in H$. Since $|J \cup \{p\}| > |I|$ and $I \cup \{p\} \notin H$, it follows from (EP') that $I \cup \{j\} \in H$ for some $j \in J \setminus I$. Hence $I \cup \{j\} \in H \cap 2^X$ and $I \cup \{j, p\} \notin H$, contradicting the maximality of I. Thus $J \cup \{p\} \notin H$ and so

$$X\delta \subseteq X \cup \{p \in V \setminus X \mid J \cup \{p\} \notin H \text{ for every } J \in H(X)\}.$$

The opposite inclusion being trivial, it follows that (4.2) holds.

Since $X \subseteq X\delta \subseteq \operatorname{Cl} X$ by part (i), it suffices to show that $X\delta$ is closed. Let $K \in H \cap 2^{X\delta}$ and $p \in V \setminus X\delta$. We must show that $K \cup \{p\} \in H$. Suppose not. Since we may replace K by any $K' \in H \cap 2^{X\delta}$ containing K, we may assume that K is maximal with respect to the properties $K \in H \cap 2^{X\delta}$ and $K \cup \{p\} \notin H$. Suppose that $K \notin H(X\delta)$. Take $K' \in H(X\delta)$. Since $|K'| > |K|$, by (EP') there exists some $k' \in K' \setminus K$ such that $K \cup \{k'\} \in H$. Hence $K \cup \{k'\} \in H \cap 2^{X\delta}$ and $K \cup \{k', p\} \notin H$, contradicting the maximality of K. Thus $K \in H(X\delta)$.

On the other hand, by (4.2), $p \notin X\delta$ implies that $p \notin X$ and $J \cup \{p\} \in H$ for some $J \in H(X)$. Suppose that $|K| > |J|$. By (EP'), we have $J \cup \{k\} \in H$ for some $k \in K \setminus J$. Since $J \in H(X)$, we get $k \notin X$ and so $k \in X\delta \setminus X$. Thus (4.2) yields $J \cup \{k\} \notin H$, a contradiction. Thus $|K| \le |J|$ and so $|K| < |J \cup \{p\}|$. Since $K \cup \{p\} \notin H$, by (EP') we get $K \cup \{j\} \in H$ for some $j \in J \setminus K$, contradicting $K \in H(Y)$. Therefore $K \cup \{p\} \in H$ and so $X\delta$ is closed as required. $\square$

The next example shows that Proposition 4.2.5(ii) cannot be generalized to arbitrary simplicial complexes (even if they are boolean representable, cf. Example 5.2.11(iii)):

Example 4.2.6. Let $\mathcal{H} = (V, H)$ with $V = \{1, \dots, 4\}$ and $H = P_{\le 2}(V) \cup \{123, 124\}$. Then $\operatorname{Cl} X = X\delta$ fails for $X = 13$.

Indeed, in this case we get $X\delta = 134$ which is not closed since $34 \in H \cap 2^{X\delta}$ but $234 \notin H$.

Chapter 5
Boolean Representations

In this chapter we begin the main subject of this monograph, boolean representations of simplicial complexes. In view of the correspondences established in Sects. 3.4 and 3.5, lattices play a major role.

5.1 Superboolean and Boolean Representations

Given $M \in M_{m \times n}(\mathbb{SB})$, the subsets of independent column vectors of M, as defined in Sect. 2.2, include the empty subset and are closed under taking subsets, and constitute therefore an example of a simplicial complex.

Given a simplicial complex $\mathcal{H} = (V, H)$, a *superboolean representation* of $\mathcal{H}$ is a superboolean matrix M with column space V such that a subset $X \subseteq V$ of column vectors of M is independent (over $\mathbb{SB}$) if and only if $X \in H$. The following theorem by Izhakian and Rhodes shows that the above example is indeed very important:

Theorem 5.1.1 ([28, Theorem 4.6]). *Every simplicial complex admits a superboolean representation.*

Proof. Let $\mathcal{H} = (V, H)$ and assume that $V = \{1, \ldots, m\}$, with the usual ordering. Let $H' = H \setminus \{\emptyset\}$. We define an $H' \times V$ superboolean matrix $M = (m_{Xp})$ by

$$m_{Xp} = \begin{cases} 1 & \text{if } p = \min X \\ 0 & \text{if } p \in X \setminus \{\min X\} \\ 2 & \text{if } p \notin X \end{cases}$$

We claim that M is a superboolean representation of $\mathcal{H}$.

Let $X \subseteq V$. We may assume that $X \neq \emptyset$. We write $X = x_1 \ldots x_n$ with $x_1 < \ldots < x_n$.

© Springer International Publishing Switzerland 2015
J. Rhodes, P.V. Silva, *Boolean Representations of Simplicial Complexes and Matroids*, Springer Monographs in Mathematics,
DOI 10.1007/978-3-319-15114-4_5

Assume first that $X \in H$. Write $X_i = x_i \ldots x_n$ for every $i \in \{1, \ldots, n\}$. Then $X_1, \ldots, X_n \in H'$ and it follows from the definition that $M[X_1, \ldots, X_n; X]$ is a lower unitriangular submatrix of M. Hence X has a witness in M and so the column vectors $M[\_, x_i]$ ($i \in \{1, \ldots, n\}$) are independent by Proposition 2.2.6.

Conversely, assume that the column vectors $M[\_, x_i]$ ($i \in \{1, \ldots, n\}$) are independent. By Proposition 2.2.6, there exists some $Y \subseteq H'$ such that $M[Y, X]$ is nonsingular. It follows that $M[Y, X]$ is congruent to a lower unitriangular matrix and has therefore a marker. If the row $M[X', X]$ is a marker, then $X \subseteq X' \in H'$ and so $X \in H$ as required. $\square$

Thus superboolean representations cover the full range of simplicial complexes. However, we can get more interesting properties by restricting the type of matrices which we are willing to admit.

Given a simplicial complex $\mathcal{H} = (V, H)$, a *boolean representation* of $\mathcal{H}$ is a boolean matrix M with column space V such that a subset $X \subseteq V$ of column vectors of M is independent (over $\mathbb{SB}$) if and only if $X \in H$. Note that independence is considered in the superboolean semiring $\mathbb{SB}$, not in the boolean semiring $\mathbb{B}$!

Similarly to the case of posets, we say that $X \subseteq V$ is *c-independent* (with respect to some fixed matrix M with column space V) if the column vectors $M[\_, x]$ ($x \in V$) are independent. Therefore a boolean representation of $\mathcal{H}$ is a boolean matrix with column space V for which X is c-independent if and only if $X \in H$.

Obviously, we can always assume that the rows in such a matrix are distinct: the representation is then said to be *reduced*. Note also that by permuting rows in a reduced representation of $\mathcal{H}$ we get an alternative reduced representation of $\mathcal{H}$. The number of rows in a boolean representation M of $\mathcal{H}$ is said to be its *degree* and is denoted by $\deg M$. We denote by $\mathrm{mindeg}\,\mathcal{H}$ the minimum degree of a boolean representation of $\mathcal{H}$.

A simplicial complex is *boolean representable* if it admits a boolean representation.

The following result, due to Izhakian and Rhodes, implies that not all simplicial complexes are boolean representable:

Proposition 5.1.2 ([28, Theorem 5.3]). *Every boolean representable simplicial complex satisfies (PR).*

Proof. Let $\mathcal{H} = (V, H)$ be a simplicial complex with a boolean representation M. Let $I, \{p\} \in H \setminus \{\emptyset\}$. By Proposition 2.2.6, there exists some nonsingular submatrix $M[J, I]$. Permuting rows and columns if necessary, we may assume that $M[J, I]$ is lower unitriangular. Let $M[j_1, \ldots, j_n; i_1, \ldots, i_n]$ be the reordered matrix.

Since $\{p\} \in H$, the column vector $M[\_, p]$ is independent. By Lemma 2.2.4, it must be nonzero. We consider now two cases.

Assume first that $M[J, p]$ is not nonzero. Let

$$r = \min\{k \in \{1, \ldots, n\} \mid m_{j_k p} = 1\}.$$

We claim that $(I \setminus \{i_r\}) \cup \{p\} \in H$. Indeed, it suffices to show that the matrix obtained by replacing the i_r column in $M[J, I]$ by the p column is still lower unitriangular. This is equivalent to saying that $m_{j_1 p} = \ldots = m_{j_{r-1} p} = 0$ and $m_{j_r p} = 1$, and all these equalities follow from the definition of r.

We may therefore assume that $M[J, p]$ contains only zeroes. Since $M[\_, p]$ is nonzero, there exists some entry $m_{qp} = 1$. It is immediate that the matrix $M[j_1, \ldots, j_{n-1}, q; i_1, \ldots, i_{n-1}, p]$ is still lower unitriangular. Thus $(I \setminus \{i_n\}) \cup \{p\} \in H$ and $\mathcal{H}$ satisfies (PR). $\square$

We shall see in Example 5.2.12 that not every simplicial complex satisfying (PR) is boolean representable.

Using Proposition 5.1.2, we produce a first example of a non boolean representable simplicial complex:

Example 5.1.3. Let $\mathcal{H} = (V, H)$ with $V = \{1, \ldots, 4\}$ and $H = P_{\leq 2}(V) \cup \{123\}$. Then $\mathcal{H}$ is not boolean representable.

Indeed, $\mathcal{H}$ fails (PR) for $123, \{4\} \in H$.

In the above example, $\mathcal{H}$ is a paving simplicial complex of dimension 2. For every $d \geq 2$, we denote by $\mathrm{Pav}(d)$ the class of all paving simplicial complexes of dimension d.

Boolean representable paving simplicial complexes play a major part in this monograph. We denote by $\mathrm{BPav}(d)$ the class of all boolean representable paving simplicial complexes of dimension d.

5.2 The Canonical Boolean Representation

We show in this section that, whenever a simplicial complex is boolean representable, the lattice of flats provides a canonical representation.

Given an $R \times V$ boolean matrix $M = (m_{rx})$ and $r \in R$, we recall the notation $Z_r = \{x \in V \mid m_{rx} = 0\}$ introduced in Sect. 3.4.

Lemma 5.2.1. *Let $\mathcal{H} = (V, H)$ be a simplicial complex and let M be an $R \times V$ boolean matrix. If M is a boolean representation of $\mathcal{H}$, then $Z_r \in \mathrm{Fl}\mathcal{H}$ for every $r \in R$.*

Proof. Let $r \in R$. Let $I \in H \cap 2^{Z_r}$ and $p \in V \setminus Z_r$. Since $M = (m_{rx})$ is a boolean representation of $\mathcal{H}$, then I is c-independent and so by Proposition 2.2.6, there exists some lower unitriangular submatrix $M[J, I]$, for a suitable ordering of I and J. Since $I \subseteq Z_r$, the row vector $M[r, I]$ contains only zeros. On the other hand, since $p \notin Z_r$, we have $m_{rp} = 1$ and so $M[\{r\} \cup J, \{p\} \cup I]$ is also lower unitriangular and therefore nonsingular. Thus $I \cup \{p\}$ is c-independent. Since M is a boolean representation of $\mathcal{H}$, it follows that $I \cup \{p\} \in H$ and so $Z_r \in \mathrm{Fl}\,\mathcal{H}$. $\square$

Let $\mathcal{H} = (V, H)$ be a simplicial complex. In view of Proposition 4.2.2(ii), we may view (Fl $\mathcal{H}, \subseteq$) as a lattice with $(X \wedge Y) = X \cap Y$ and the determined join

$$(X \vee Y) = \cap\{Z \in \text{Fl } \mathcal{H} \mid X \cup Y \subseteq Z\} = \text{Cl}(X \cup Y).$$

Let

$$V^{\iota} = \{x \in V \mid \{x\} \in H\}.$$

The top element of Fl $\mathcal{H}$ is V by Proposition 4.2.2(i). We compute now its bottom element.

Proposition 5.2.2. *Let $\mathcal{H} = (V, H)$ be a simplicial complex. Then:*

(i) $\cap(\text{Fl } \mathcal{H}) = V \setminus V^{\iota}$;
(ii) If $\mathcal{H}$ is trim, then $\cap(\text{Fl } \mathcal{H}) = \emptyset$.

Proof. (i) Write $X = V \setminus V^{\iota}$. Let $I \in H \cap 2^X$ and $p \in V \setminus X$. It follows that $I = \emptyset$ and $p \in V^{\iota}$, hence $I \cup \{p\} \in H$ and X is closed.

Now let $Y \in \text{Fl } \mathcal{H}$ be arbitrary. Suppose that $p \in X \setminus Y$. Since $\emptyset \in H \cap 2^Y$, we must have $\emptyset \cup \{p\} \in H$, contradicting $p \notin V^{\iota}$. Hence $X \subseteq Y$ and so $X = \cap(\text{Fl } \mathcal{H})$.

(ii) If $\mathcal{H}$ is trim, then $V^{\iota} = V$. $\square$

For every simplicial complex $\mathcal{H} = (V, H)$, we define a (Fl $\mathcal{H}$) $\times V$ boolean matrix Mat $\mathcal{H} = (m_{Xp})$ by

$$m_{Xp} = \begin{cases} 0 & \text{if } p \in X \\ 1 & \text{otherwise} \end{cases}$$

If $\mathcal{H}$ is simple, we may identify $x \in V$ with $\{x\}$ and V with $P_1(V)$, and then it follows from Proposition 4.2.2(iv) that $V \subseteq \text{Fl } \mathcal{H}$. Moreover, we may write (Fl $\mathcal{H}, V$) $\in$ FLg. Indeed, for every $X \in \text{Fl } \mathcal{H}$, we have $X = \vee\{\{x\} \mid x \in X\}$. Thus, if $\mathcal{H}$ is simple, Mat $\mathcal{H}$ coincides with the matrix $M(\text{Fl } \mathcal{H}, V)$ defined in Sect. 3.3 as the boolean representation of a $\vee$-generated lattice. We shall see in Theorem 5.2.5 that Mat $\mathcal{H}$ constitutes somehow the *canonical* unique biggest boolean representation for $\mathcal{H}$ (when $\mathcal{H}$ is boolean representable). However, there exist many smaller boolean representations, even in the matroid case.

Lemma 5.2.3. *Let $\mathcal{H} = (V, H)$ be a simplicial complex and let $X \subseteq V$ be c-independent with respect to Mat $\mathcal{H}$. Then $X \in H$.*

Proof. We use induction on $|X|$. The case $|X| = 0$ being trivial, we assume that $|X| = m > 0$ and the claim holds for $|X| = m - 1$.

By Proposition 2.2.6, X has a witness Y in $M = M(\text{Fl } \mathcal{H}, V)$. We may assume that $X = \{x_1, \ldots, x_m\}$, $Y = \{Y_1, \ldots, Y_m\}$ and $M[Y, X]$ is lower unitriangular, with the rows (respectively the columns) ordered by $Y_1, \ldots, Y_m$ (respectively $x_1, \ldots, x_m$). The first row yields $x_1 \notin Y_1$ and $x_2, \ldots, x_m \in Y_1$.

Now, since $X' = \{x_2, \ldots, x_m\}$ is c-independent, it follows from the induction hypothesis that $X' \in H$. Together with $X' \subseteq Y_1 \in \text{Fl } \mathcal{H}$ and $x_1 \notin Y_1$, this yields $X = X' \cup \{x_1\} \in H$ as required. $\square$

Given matrices M_1 and M_2 with the same number of columns, we define $M_1 \oplus_b M_2$ to be the matrix obtained by concatenating the matrices M_1 and M_2 by

$$\binom{M_1}{M_2}$$

and removing repeated rows (leaving only the first occurrence from top to bottom, say). We refer to this matrix as M_1 *stacked over* M_2.

The following proposition gives a first glimpse of how we may operate within the (finite) set of reduced boolean representations of $\mathcal{H}$:

Proposition 5.2.4. *Let $\mathcal{H} = (V, H)$ be a simplicial complex.*

(i) If M_1 and M_2 are reduced boolean representations of $\mathcal{H}$, so is $M_1 \oplus_b M_2$.
(ii) If M is a reduced boolean representation of $\mathcal{H}$ and we add/erase a row which is the sum of other rows in $\mathbb{B}^{|V|}$, we get a matrix M' which is also a reduced boolean representation of $\mathcal{H}$.

Proof. (i) Since M_1 and M_2 have both column space V, the matrix $M = M_1 \oplus_b M_2$ is well defined and has no repeated rows by definition. Let R be the row space of M and let $X \subseteq V$. We show that

$$X \text{ is c-independent with respect to } M \iff X \in H \qquad (5.1)$$

by induction on $|X|$. The case $|X| = 0$ being trivial, assume that $|X| > 0$ and (5.1) holds for smaller values of $|X|$.

Suppose that X is c-independent with respect to M. By permuting rows of $M_1 \oplus_b M_2$ if necessary, and using the appropriate ordering of V, we may assume that there exists some $P \subseteq R$ such that $M[P, X]$ is lower unitriangular. Let p_1 (respectively x_1) denote the first element of P (respectively X) for these orderings, so $M[P \setminus \{p_1\}, X \setminus \{x_1\}]$ is the submatrix of $M[P, X]$ obtained by deleting the first row and the first column. Since reduced boolean representations are closed under permuting rows, we may assume without loss of generality that the row $M[p_1, V]$ came from the matrix M_1. On the other hand, since $X \setminus \{x_1\}$ is c-independent with respect to M, it follows from the induction hypothesis that $X \setminus \{x_1\} \in H$ and so (since M_1 is a boolean representation of $\mathcal{H}$) $X \setminus \{x_1\}$ is c-independent with respect to M_1. Hence M_1 has a singular submatrix of the form $M_1[P', X \setminus \{x_1\}]$. Now $M_1[P' \cup \{p_1\}, X]$ is still a nonsingular matrix because the unique nonzero entry in the row $M_1[p_1, X]$ is $M_1[p_1, x_1]$. Hence X is c-independent with respect to M_1 and so $X \in H$.

Conversely, if $X \in H$, then X is c-independent with respect to M_1 and so X is c-independent with respect to M by Lemma 2.2.3(i). Thus (5.1) holds and so M is a reduced boolean representation of $\mathcal{H}$ as claimed.

(ii) By Corollary 2.2.7. $\square$

Proposition 5.2.4(i) immediately implies that if $\mathcal{H}$ admits a reduced boolean representation, then there exists a unique maximal one (up to permuting rows). The main theorem of this section provides a more concrete characterization.

Theorem 5.2.5. *Let* $\mathcal{H} = (V, H)$ *be a simplicial complex. Then the following conditions are equivalent:*

(i) $\mathcal{H}$ has a boolean representation;
(ii) Mat $\mathcal{H}$ is a reduced boolean representation of $\mathcal{H}$.

Moreover, in this case any other reduced boolean representation of $\mathcal{H}$ is congruent to a submatrix of Mat $\mathcal{H}$.

Proof. (i) $\Rightarrow$ (ii). Write $M = $ Mat $\mathcal{H}$. Suppose that $\mathcal{H}$ has a boolean representation $N = (n_{rx})$. Then we may assume that the $R \times V$ matrix N is reduced. By Lemma 5.2.1, we have $Z_r \in$ Fl $\mathcal{H}$ for every $r \in R$. For every $x \in V$, we have

$$n_{rx} = 0 \Leftrightarrow x \in Z_r \Leftrightarrow \{x\} \subseteq Z_r \Leftrightarrow M[Z_r, x] = 0,$$

hence N is (up to permutation of rows) a submatrix of M.

We claim that M is also a boolean representation of $\mathcal{H}$. Indeed, let $X \subseteq V$. If $X \in H$, then X is c-independent with respect to N since N is a boolean representation of $\mathcal{H}$, hence X is c-independent with respect to M by Lemma 2.2.3(i). The converse implication follows from Lemma 5.2.3, hence M is a boolean representation of $\mathcal{H}$. Finally, it follows from its definition that M is reduced.

(ii) $\Rightarrow$ (i). Trivial. $\square$

In the next theorem, we explore the potential of Mat $\mathcal{H}$ to characterize the faces of a simplicial complex:

Theorem 5.2.6. *Let* $\mathcal{H} = (V, H)$ *be a boolean representable simplicial complex and let $X \subseteq V$. Then the following conditions are equivalent:*

(i) $X \in H$;
(ii) X is a transversal of the successive differences for some chain of Fl $\mathcal{H}$;
(iii) X is a partial transversal of the successive differences for some maximal chain of Fl $\mathcal{H}$;
(iv) X admits an enumeration $x_1, \ldots, x_k$ such that

$$\mathrm{Cl}(x_1, \ldots, x_k) \supset \mathrm{Cl}(x_2, \ldots, x_k) \supset \ldots \supset \mathrm{Cl}(x_k) \supset \mathrm{Cl}(\emptyset);$$

(v) X admits an enumeration $x_1, \ldots, x_k$ such that

$$x_i \notin \mathrm{Cl}(x_{i+1}, \ldots, x_k) \quad \text{for } i = 1, \ldots, k.$$

Proof. (i) $\Rightarrow$ (ii). Assume $X \in H$. By Theorem 5.2.5, X is c-independent with respect to $M = \mathrm{Mat}\,\mathcal{H}$. By Proposition 2.2.6, we may assume that M has a lower unitriangular submatrix $M[Y, X]$. Let $Y = \{Z_1, \ldots, Z_k\}$ and $X = \{x_1, \ldots, x_k\}$ be the corresponding enumerations of Y and X. For $i \in \{1, \ldots, k\}$, let $Z_i' = Z_1 \cap \ldots \cap Z_i$. By Proposition 4.2.2(ii), we have $Z_i' \in \mathrm{Fl}\,\mathcal{H}$. Moreover, $x_i \in Z_{i-1}' \setminus Z_i'$ for $i > 1$ and $x_1 \in V \setminus Z_1'$, hence X is a transversal of the successive differences for the chain

$$V \supset Z_1' \supset Z_2' \supset \ldots \supset Z_{k-1}' \supset Z_k'$$

of $\mathrm{Fl}\,\mathcal{H}$.

(ii) $\Rightarrow$ (i). Assume that X is a transversal of successive differences for the chain

$$Z_0 \supset Z_1 \supset Z_2 \supset \ldots \supset Z_k$$

of $\mathrm{Fl}\,\mathcal{H}$. Then X admits an enumeration $x_1, \ldots, x_k$ such that $x_i \in Z_{i-1} \setminus Z_i$ for every $i \in \{1, \ldots, k\}$. It follows easily that $M[Z_1, \ldots, Z_k; x_1, \ldots, x_k]$ is lower unitriangular, hence X is c-independent with respect to M and therefore $X \in H$ by Lemma 5.2.3.

(ii) $\Leftrightarrow$ (iv). Assume that X is a transversal for the successive differences of the chain $F_0 \supset \ldots \supset F_k$ in $\mathrm{Fl}\,\mathcal{H}$. Take $x_i \in X \cap (F_{i-1} \setminus F_i)$ for $i \in \{1, \ldots, k\}$. Then $\mathrm{Cl}(x_{i+1}, \ldots, x_k) \subseteq F_i$ for every $i \in \{1, \ldots, k\}$ yields $\mathrm{Cl}(x_i, \ldots, x_k) \supset \mathrm{Cl}(x_{i+1}, \ldots, x_k)$.

(v) $\Leftrightarrow$ (iv) $\Rightarrow$ (ii) $\Leftrightarrow$ (iii). Straightforward. $\square$

We can now prove another characterization of boolean representability:

Corollary 5.2.7. *Let $\mathcal{H} = (V, H)$ be a simplicial complex. Then the following conditions are equivalent:*

(i) $\mathcal{H}$ is boolean representable;
(ii) Every $X \in H$ admits an enumeration $x_1, \ldots, x_k$ satisfying

$$\mathrm{Cl}(x_1, \ldots, x_k) \supset \mathrm{Cl}(x_2, \ldots, x_k) \supset \ldots \supset \mathrm{Cl}(x_k) \supset \mathrm{Cl}(\emptyset); \qquad (5.2)$$

(iii) For every nonempty $X \in H$, there exists some $x \in X$ such that $x \notin \mathrm{Cl}(X \setminus \{x\})$.

Proof. (i) $\Rightarrow$ (iii). By Proposition 5.2.6.

(iii) $\Rightarrow$ (ii). By induction on $|X|$.

(ii) $\Rightarrow$ (i). Let $X \in H$. In view of (5.2), we can use the flats $\mathrm{Cl}\,(x_i, \ldots, x_k)$ as a witness for X, hence X is c-independent with respect to Mat $\mathcal{H}$. Lemma 5.2.3 yields the reciprocal implication and so Mat $\mathcal{H}$ provides a boolean representation of $\mathcal{H}$. $\square$

Other immediate consequences of Theorem 5.2.6 are the following:

Corollary 5.2.8. *Let $\mathcal{H}$ be a boolean representable simplicial complex of dimension d. Then* $\mathrm{ht}\,\mathrm{Fl}\,\mathcal{H} = d + 1$.

Corollary 5.2.9. *A boolean representable simplicial complex is fully determined by its flats.*

The important subcase of boolean representations of matroids was studied in [28, 29] and the following fundamental result was proved for simple matroids by Izhakian and Rhodes [29, Theorem 4.1]. The general case can now be easily deduced from Corollary 5.2.7:

Theorem 5.2.10. *Let $\mathcal{H}$ be a matroid. Then* Mat $\mathcal{H}$ *is a boolean representation of $\mathcal{H}$.*

Proof. Let $X \in H$, say $X = \{x_1, \ldots, x_k\}$. We claim that (5.2) holds. Suppose that $\mathrm{Cl}\,(x_i, \ldots, x_k) = \mathrm{Cl}\,(x_{i+1}, \ldots, x_k)$ for some $i \in \{1, \ldots, k\}$. Then $x_i \in \mathrm{Cl}\,(x_{i+1}, \ldots, x_k)$ and it follows from Proposition 4.2.5 that there exists some $I \subseteq \{x_{i+1}, \ldots, x_k\}$ such that $I \in H$ and $I \cup \{x_i\} \notin H$. However, $I \cup \{x_i\} \subseteq X \in H$, a contradiction. Thus (5.2) holds and so $\mathcal{H}$ is boolean representable by Corollary 5.2.7. Therefore Mat $\mathcal{H}$ is a boolean representation of $\mathcal{H}$ by Theorem 5.2.5. $\square$

We consider next as examples the *tetrahedron complexes*. For $n \in \{0, \ldots, 4\}$, we denote by $T_n = (V, H)$ a simple simplicial complex of dimension ≤ 2 having precisely n faces of dimension 2 and satisfying $|V| = 4$. We can think of T_n as a complete graph on 4 points (the skeleton of a tetrahedron) with n triangles adjoined.

Example 5.2.11. (i) T_0, T_3 and T_4 are matroids and therefore boolean representable.
  (ii) T_1 does not satisfy (PR) and so is not boolean representable.
 (iii) T_2 is not a matroid but it is boolean representable.

Indeed, it is easy to check that T_n is a matroid for $n \in \{0, 3, 4\}$, hence T_n is boolean representable by Theorem 5.2.10. The case T_1 has already been discussed in Example 5.1.3.

For T_2, assume that $V = \{1, \ldots, 4\}$ and 123 and 124 are the 2 faces of dimension 2. Since T_2 fails (EP) for $I = 123$ and $J = 34$, then T_2 is not a matroid. However, since $\mathrm{Fl}\,T_2 = \mathcal{P}_{\leq 1}(V) \cup \{12, V\}$, it follows easily from Corollary 5.2.7 that T_2 is boolean representable.

In fact, in this case the lattice of flats can be depicted as

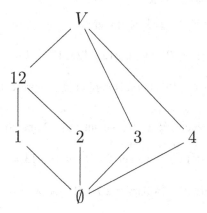

and so there exist maximal chains $\emptyset \subset 1 \subset 12 \subset V$ and $\emptyset \subset 4 \subset V$ of different length. Hence Fl T_2 does not satisfy the *Jordan-Dedekind* condition and so is not semimodular by [23, Theorem 374]. We recall that a lattice L is said to be *semimodular* if there is no sublattice of the form

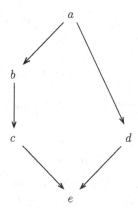

with d covering e.

On the other hand, L is *atomistic* if every element is a join of atoms (B being the join of the empty set). It is said to be *geometric* if it is both semimodular and atomistic. It is well known that a finite lattice is geometric if and only if it is isomorphic to the lattice of flats of some (finite) matroid [39, Theorem 1.7.5].

Hence the above example shows that properties such as semimodularity or the Jordan-Dedekind condition, which hold in the lattice of flats of a matroid, may fail in the lattice of flats of a boolean representable simplicial complex, even though it is simple and paving.

We end this section by providing an example of a non boolean representable simplicial complex satisfying (PR):

Example 5.2.12. Let $V = \{1, 2, 3, 4, 5\}$ and

$$H = P_{\leq 2}(V) \cup \{123, 124, 135, 145\}.$$

Then $\mathcal{H} = (V, H)$ satisfies (PR) but it is not boolean representable.

Indeed, it is straightforward to check (PR) for $\mathcal{H}$.

On the other hand, since $125 \notin H$, we get $5 \in \overline{12}$ and so $\overline{12}$ contains the facet 25, yielding $\overline{12} = V$ by Proposition 4.2.3.

Similarly, we get $4 \in \overline{13}$ and we get $4 \in \overline{23}$ and so $\overline{13} = \overline{23} = V$ since 34 is a facet.

Hence condition (iii) in Corollary 5.2.7 fails for $X = 123$, and so $\mathcal{H}$ is not boolean representable.

5.3   Low Dimensions

It is very easy to characterize boolean representability for simplicial complexes of dimension ≤ 1:

Recall the notation $V^{\iota} = V \cap H$ from Sect. 5.2. Given a simplicial complex $\mathcal{H} = (V, H)$ of dimension ≥ 0, we define a graph

$$\Gamma \mathcal{H} = (V^{\iota}, P_2(V^{\iota}) \setminus H).$$

Proposition 5.3.1. *The following conditions are equivalent for a simplicial complex $\mathcal{H}$ of dimension ≤ 1:*

 (i) $\mathcal{H}$ is boolean representable;
 (ii) The connected components of $\Gamma \mathcal{H}$ are cliques.

Proof. (i) $\Rightarrow$ (ii). Assume that $\Gamma \mathcal{H}$ contains a connected component C which is not a clique. Then there exist distinct edges a — b — c in C such that a is not adjacent to c i.e. $ac \in H$. Since $a \in H$ and $ab \notin H$, we get $b \in \overline{a}$. Since $b \in H$ and $bc \notin H$, we get $c \in \overline{a}$. Similarly, $a \in \overline{c}$ and so it follows from Corollary 5.2.7 that $\mathcal{H}$ is not boolean representable.

(ii) $\Rightarrow$ (i). Assume that the connected components of $\Gamma \mathcal{H}$ are cliques. Given $a \in V^{\iota}$, let C_a denote the vertices in the connected component of a. We claim that, for every $x \in V$,

$$\overline{x} = \begin{cases} V \setminus V^{\iota} & \text{if } x \notin V^{\iota} \\ (V \setminus V^{\iota}) \cup C_x & \text{otherwise} \end{cases}$$

Indeed, it follows from Proposition 5.2.2 that $\overline{\emptyset} = V \setminus V^\iota$, hence also $\overline{x} = V \setminus V^\iota$ when $x \in V \setminus V^\iota$. Assume now that $x \in V^\iota$. Since $xy \notin H$ for every $y \in C_x \setminus \{x\}$, we get $(V \setminus V^\iota) \cup C_x \subseteq \overline{x}$. Suppose that $p \in V \setminus ((V \setminus V^\iota) \cup C_x)$ and $I \subseteq (V \setminus V^\iota) \cup C_x$ is such that $I \in H$. We may assume that $I \neq \emptyset$ and since C_x is a clique in $\Gamma \mathcal{H}$ it follows easily that $I = y$ for some $y \in C_x$. Hence p belongs to some other connected component of $\Gamma \mathcal{H}$ and so $xy \in H$. Hence $(V \setminus V^\iota) \cup C_x$ is closed and so $\overline{x} = (V \setminus V^\iota) \cup C_x$.

Now, if $ab \in H$, then a and b belong to different connected components and so $b \notin \overline{a}$. Since $c \notin \overline{\emptyset}$ for every $c \in V^\iota$, it follows from Corollary 5.2.7 that $\mathcal{H}$ is boolean representable. $\square$

Corollary 5.3.2. *Every simplicial complex of dimension ≤ 0 is boolean representable.*

Proof. If $\dim \mathcal{H} = -1$, then (V, H) is represented by a zero row matrix.

If $\dim \mathcal{H} = 0$, then $\Gamma \mathcal{H}$ is a complete graph and $\mathcal{H}$ is boolean representable by Proposition 5.3.1. $\square$

5.4 Lattice Representations

The correspondences established in Sects. 3.4 and 3.5 between boolean matrices and $\vee$-generated lattices lead us naturally to the idea of lattice representations of a simplicial complex $\mathcal{H} = (V, H)$.

We say that $(L, V) \in \mathrm{FLg}$ is a *lattice representation* of $\mathcal{H}$ if the matrix $M(L, V)$ is a boolean representation of $\mathcal{H}$. Our main goal in this section is to relate the closure operator Cl defined in Sect. 4.2 with the closure operator induced by a representation. Given a lattice representation (L, V) of $\mathcal{H}$, we denote by Cl_L the corresponding closure operator on 2^V as defined in Sect. 3.6.

Lemma 5.4.1. *Let $\mathcal{H} = (V, H)$ be a simplicial complex and let (L, V) be a lattice representation of $\mathcal{H}$. Let $X \subseteq V$. Then:*

(i) $\mathrm{Cl}\, X \subseteq \mathrm{Cl}_L X$;
(ii) $\mathrm{Cl}\, X = \mathrm{Cl}_L X$ *if* $L = \mathrm{Fl}\mathcal{H}$.

Proof. (i) We have $\mathrm{Cl}\, X = \cap \{Z \in \mathrm{Fl}\ \mathcal{H} \mid X \subseteq Z\}$ and in view of (3.10) and Lemma 5.2.1 also

$$\mathrm{Cl}_L X = Z_{\vee_L X} \in \mathrm{Fl}\ \mathcal{H}\ .$$

Since $X \subseteq \mathrm{Cl}_L X$, we get $\mathrm{Cl}\, X \subseteq \mathrm{Cl}_L X$.
(ii) If $L = \mathrm{Fl}\ \mathcal{H}$, then $\mathrm{Cl}_L X = Z_{\vee_L X} = Z_{\mathrm{Cl}\, X} = \mathrm{Cl}\, X. \square$

Now, a straightforward adaptation of the proof of Theorem 5.2.6 yields:

Theorem 5.4.2. *Let* $\mathcal{H} = (V, H)$ *be a simplicial complex and let* (L, V) *be a lattice representation of* $\mathcal{H}$. *Let* $X \subseteq V$. *Then the following conditions are equivalent:*

(i) $X \in H$;
(ii) X *is a transversal of the successive differences for some chain of* $\mathrm{Fl}(L, V)$;
(iii) X *is a partial transversal of the successive differences for some maximal chain of* $\mathrm{Fl}(L, V)$;
(iv) X *admits an enumeration* $x_1, \ldots, x_k$ *such that*

$$\mathrm{Cl}_L(x_1, \ldots, x_k) \supset \mathrm{Cl}_L(x_2, \ldots, x_k) \supset \ldots \supset \mathrm{Cl}_L(x_k) \supset \mathrm{Cl}_L(\emptyset);$$

(v) X *admits an enumeration* $x_1, \ldots, x_k$ *such that*

$$x_i \notin \mathrm{Cl}_L(x_{i+1}, \ldots, x_k) \quad for\ i = 1, \ldots, k.$$

The next result shows how to associate a lattice representation to a matrix representation in the simple case.

Proposition 5.4.3. *Let* $\mathcal{H} = (V, H)$ *be a simple simplicial complex and let* M *be a boolean representation of* $\mathcal{H}$. *Then:*

(i) *We may view* V *as a* $\vee$-*generating set for* $\mathrm{Fl}\, M$;
(ii) $(\mathrm{Fl}\, M, V)$ *is a lattice representation of* $\mathcal{H}$.

Proof. (i) Write $M = (m_{ip})$. Since $\mathcal{H}$ is simple, all the columns of M are distinct and nonzero. By Proposition 3.4.1, we have $(\mathrm{Fl}\, M, \mathcal{Y}(M)) \in \mathrm{FLg}$, where $\mathcal{Y}(M) = \{Y_p \mid p \in V\}$ and $Y_p = \cap\{Z_i \mid m_{ip} = 0\}$.

Assume that $p, q \in V$ are distinct. Then $m_{jp} \neq m_{jq}$ for some j. Assuming without loss of generality $m_{jp} = 0$, it follows that $j \in Y_p \setminus Y_q$ and so $Y_p \neq Y_q$. Therefore we can identify V with the $\vee$-generating set $\mathcal{Y}(M) \subseteq \mathrm{Fl}\, M$.

(ii) Let $M' = M(\mathrm{Fl}\, M, V)$. Then M' can be obtained from M by successively inserting a row of zeroes (corresponding to the fact that $V \in \mathrm{Fl}\, M$) and sums of rows of M in $\mathbb{B}^{|V|}$ (corresponding to intersections of the Z_i). Since we may assume M to be reduced, the claim follows now from Proposition 5.2.4(ii). $\square$

5.5   The Lattice of Lattice Representations

We show in this section that we can organize the lattice representations of a simple simplicial complex into a lattice of their own.

We define a quasi-order on FLg by $(L, A) \geq (L', A)$ if there exists some $\vee$-map $\varphi : L \to L'$ such that $\varphi|_A = 1_A$. Note that such a φ is necessarily onto: if $x' \in L'$, we may write $x' = (a_1 \vee_{L'} \ldots \vee_{L'} a_k)$ for some $a_1, \ldots, a_k \in A$, hence

$$x' = (a_1 \vee_{L'} \ldots \vee_{L'} a_k) = (a_1\varphi \vee_{L'} \ldots \vee_{L'} a_k\varphi)$$
$$= (a_1 \vee_L \ldots \vee_L a_k)\varphi \in L\varphi.$$

Recalling Corollary 3.5.3, we prove next the following alternative characterization of the above quasi-order:

Proposition 5.5.1. *Let* $(L, A), (L', A) \in$ FLg. *Then the following conditions are equivalent:*

(i) $(L', A) \le (L, A)$;
(ii) $\mathrm{Fl}(L', A) \subseteq \mathrm{Fl}(L, A)$.

Proof. (i) $\Rightarrow$ (ii). There exists some (onto) $\vee$-map $\varphi : L \to L'$ such that $\varphi|_A = 1_A$. We claim that

$$Z_{x'} = Z_{\vee(x'\varphi^{-1})} \qquad (5.3)$$

holds for every $x' \in L'$. Let $a \in Z_{x'}$. Then $a \le x'$. Write $x = \vee(x'\varphi^{-1})$. Since φ is a $\vee$-map, we have $x\varphi = x'$. Moreover, $(x \vee_L a)\varphi = (x\varphi \vee_{L'} a\varphi) = (x' \vee_{L'} a) = x'$, hence $(x \vee_L a) \in x'\varphi^{-1}$ and so $(x \vee_L a) \le \max(x'\varphi^{-1}) = x$. Thus $a \le x$ and so $Z_{x'} \subseteq Z_x = Z_{\vee(x'\varphi^{-1})}$.

Conversely, assume that $a \in Z_x$. Then $a \le \vee(x'\varphi^{-1})$ and so $a = a\varphi \le x'$. Hence $Z_x \subseteq Z_{x'}$ and so (5.3) holds. Therefore $\mathrm{Fl}(L', A) \subseteq \mathrm{Fl}(L, A)$.

(ii) $\Rightarrow$ (i). We define a map $\varphi : L \to L'$ as follows. For every $x \in L$, let $x\varphi = \vee_{L'} Z_x$.

For every $a \in A$, we have $a\varphi = \vee_{L'} \{b \in A \mid b \le_L a\}$. Since $a \le_L a$, we get $a \le_{L'} a\varphi$. Write $Z_a = \{b \in A \mid b \le_L a\}$ and $Z'_a = \{b \in A \mid b \le_{L'} a\}$. We have $Z'_a = Z_x$ for some $x \in L$ since $\mathrm{Fl}(L', A) \subseteq \mathrm{Fl}(L, A)$. It follows that $a \in Z_x$ and so $Z_a \subseteq Z_x = Z'_a$. By Lemma 3.5.1(ii), we get $a\varphi = \vee_{L'} Z_a \le_{L'} \vee_{L'} Z'_a = a$ and so $a\varphi = a$.

In view of Lemma 3.5.1(iii), φ is order-preserving. Hence, given $x_1, x_2 \in L$, we have $x_i\varphi \le (x_1 \vee x_2)\varphi$ for $i = 1, 2$ and so

$$(x_1\varphi \vee x_2\varphi) \le (x_1 \vee x_2)\varphi. \qquad (5.4)$$

Since $\mathrm{Fl}(L', A) \subseteq \mathrm{Fl}(L, A)$, we have $Z_{x_1\varphi \vee x_2\varphi} = Z_y$ for some $y \in L$. We claim that

$$(x_1 \vee x_2) \le y. \qquad (5.5)$$

Let $a \in Z_{x_i}$. Since φ is order-preserving, we get

$$a = a\varphi \le x_i\varphi \le (x_1\varphi \vee x_2\varphi),$$

hence $Z_{x_i} \subseteq Z_{x_1\varphi \vee x_2\varphi} = Z_y$ and so $x_i \le y$ by Lemma 3.5.1(iii). Thus (5.5) holds.
Now by Lemma 3.5.1 and (5.5) we get

$$(x_1 \vee x_2)\varphi = \vee_{L'} Z_{x_1 \vee x_2} \le \vee_{L'} Z_y = \vee_{L'} Z_{x_1\varphi \vee x_2\varphi} = (x_1\varphi \vee x_2\varphi).$$

Together with (5.4), this implies that φ is a $\vee$-morphism. Since $B\varphi = B$, φ is actually a $\vee$-map and so $(L', A) \leq (L, A)$. $\square$

Recall that, if $A' = \{Z_a \mid a \in A\}$, it follows from Proposition 3.5.2 that $(\mathrm{Fl}(L, A), A') \cong (L, A)$ holds for every $(L, A) \in \mathrm{FLg}$. We identify A' with A to simplify notation.

Let $\mathcal{H} = (V, H)$ be a simple simplicial complex. We denote by $\mathrm{LR}\,\mathcal{H}$ the class of all lattice representations of $\mathcal{H}$. We restrict to $\mathrm{LR}\,\mathcal{H}$ the quasi-order previously defined on FLg. Note that $(\mathrm{Fl}\,\mathcal{H}, V) \in \mathrm{LR}\,\mathcal{H}$ by Proposition 5.4.3.

If $(L, V) \in \mathrm{LR}\,\mathcal{H}$ and $M = M(L, V)$, then by Lemma 5.2.1 we have $Z_x \in \mathrm{Fl}\,\mathcal{H}$ for every $x \in L$. Thus Proposition 5.5.1 yields $(\mathrm{Fl}\,\mathcal{H}, V) \geq (L, V)$.

If we consider the set of full $\wedge$-subsemilattices $\mathrm{FSub}_\wedge \mathrm{Fl}\,\mathcal{H}$ ordered by inclusion, we have a poset closed under intersection, hence a $\wedge$-semilattice and therefore a lattice with the determined join

$$(F_1 \vee F_2) = \cap\{F \in \mathrm{FSub}_\wedge \mathrm{Fl}\,\mathcal{H} \mid F_1 \cup F_2 \subseteq F\}.$$

It is easy to check that

$$\theta : (\mathrm{LR}\,\mathcal{H}, \leq) \to (\mathrm{FSub}_\wedge \mathrm{Fl}\,\mathcal{H}, \leq)$$
$$(L, V) \mapsto \mathrm{Fl}(L, V)$$

is a well-defined map. Indeed, let $(L, V) \in \mathrm{LR}\,\mathcal{H}$. Then $\mathrm{Fl}(L, V) \subseteq \mathrm{Fl}\,\mathcal{H}$ by Lemma 5.2.1, and it is closed under intersection by (3.9). Note that $(L, V) \in \mathrm{FLg}$ implies $V \subseteq L \setminus \{B\}$. Since $\emptyset = Z_B$ and $V = Z_T$, it follows that $\mathrm{Fl}(L, V) \in \mathrm{FSub}_\wedge \mathrm{Fl}\,\mathcal{H}$ and so θ is a well-defined map.

Our next goal is to build an isomorphism from θ. A first obstacle is the fact that θ is not onto: not every $F \in \mathrm{FSub}_\wedge \mathrm{Fl}\,\mathcal{H}$ is rich enough to represent $\mathcal{H}$.

Before characterizing the image of θ, it is convenient to characterize which lattices provide boolean representations.

Proposition 5.5.2. *Let $\mathcal{H} = (V, H)$ be a boolean representable simple simplicial complex and let $(\mathrm{Fl}\,\mathcal{H}, V) \geq (L, V) \in \mathrm{FLg}$. Then the following conditions are equivalent:*

(i) $(L, V) \in \mathrm{LR}\,\mathcal{H}$;
(ii) Every $X \in H$ admits an enumeration $x_1, \ldots, x_k$ satisfying

$$\mathrm{Cl}_L(x_1, \ldots, x_k) \supset \mathrm{Cl}_L(x_2, \ldots, x_k) \supset \ldots \supset \mathrm{Cl}_L(x_k) \supset \mathrm{Cl}_L(\emptyset).$$

Proof. (i) $\Rightarrow$ (ii). By Proposition 5.4.2.

(ii) $\Rightarrow$ (i). Let $X \subseteq V$. We must show that $X \in H$ if and only if X is c-independent with respect to $M(L, V)$.

Assume that $X \in H$. Since $\mathrm{Cl}_L Y = V \cap (\vee_L Y) \downarrow = Z_{\vee_L Y}$ for every $Y \subseteq V$ by (3.10), it follows from (ii) that the rows $\mathrm{Cl}_L(x_i, \ldots, x_k)$ act as a witness for X in $M(L, V)$ and so X is c-independent.

Conversely, assume that X is c-independent. By Proposition 5.5.1, we have $\mathrm{Fl}(L, V) \subseteq \mathrm{Fl}\,\mathcal{H}$ and so $M(L, V)$ is a submatrix of $M(\mathrm{Fl}\,\mathcal{H}, V)$. Hence X is c-independent with respect to $M(\mathrm{Fl}(V, H), V)$ by Lemma 2.2.3(i) and so $X \in H$ by Lemma 5.2.3. $\square$

To characterize the image of θ, we associate a closure operator on 2^V to every $F \in \mathrm{FSub}_\wedge \mathrm{Fl}\,\mathcal{H}$ by

$$\mathrm{Cl}_F X = \cap\{Z \in F \mid X \subseteq Z\}.$$

(i.e. $\mathrm{Cl}_F = F\Phi'$, see Proposition A.2.4 in the Appendix). Note also that F is a $\wedge$-semilattice and therefore a lattice with the determined join.

Corollary 5.5.3. *Let* $\mathcal{H} = (V, H)$ *be a boolean representable simple simplicial complex and let* $F \in \mathrm{FSub}_\wedge \mathrm{Fl}\,\mathcal{H}$. *Then the following conditions are equivalent:*

(i) $F \in \mathrm{Im}\,\theta$;
(ii) *Every* $X \in H$ *admits an enumeration* $x_1, \ldots, x_k$ *satisfying*

$$\mathrm{Cl}_F(x_1, \ldots, x_k) \supset \mathrm{Cl}_F(x_2, \ldots, x_k) \supset \ldots \supset \mathrm{Cl}_F(x_k) \supset \mathrm{Cl}_F(\emptyset).$$

Proof. (i) $\Rightarrow$ (ii). Assume that $F = \mathrm{Fl}(L, V)$ for some $(L, V) \in \mathrm{LR}\,\mathcal{H}$. Then $\mathrm{Cl}_F = \mathrm{Cl}_L$ and (ii) follows from Proposition 5.5.2.

(ii) $\Rightarrow$ (i). As noted before, since F is a $\cap$-subsemilattice of $\mathrm{Fl}\,\mathcal{H}$, it constitutes a lattice of its own with intersection as meet and the determined join

$$(X \vee Y) = \mathrm{Cl}_F(X \cup Y) \quad (X, Y \in F).$$

Identifying V with $\{\mathrm{Cl}_F\{x\} \mid x \in V\}$ (note that the closures are distinct in view of (ii) and $\mathcal{H}$ being simple), we can take $L = F$ to define $(L, V) \in \mathrm{FLg}$. Now, in view of Proposition 3.5.2(ii), Cl_F coincides with the closure Cl_L, and so $(L, V) \in \mathrm{LR}\,\mathcal{H}$ by Proposition 5.5.2. Therefore $F \in \mathrm{Im}\,\theta$. $\square$

Next, we claim that

$$\mathrm{FSub}_\wedge \mathrm{Fl}\,\mathcal{H} \setminus \mathrm{Im}\,\theta \text{ is a down set of } \mathrm{FSub}_\wedge \mathrm{Fl}\,\mathcal{H} . \tag{5.6}$$

Indeed, every $F \in \mathrm{FSub}_\wedge \mathrm{Fl}\,\mathcal{H}$, being a $\cap$-subsemilattice of $\mathrm{Fl}\,\mathcal{H}$, constitutes a lattice of its own right with the determined join. In view of Proposition 2.2.6, the condition $F \in \mathrm{Im}\,\theta$ reduces to whether the matrix $M(F, V)$ produces enough witnesses to recognize all the faces of $\mathcal{H}$. Therefore, if $F' \supseteq F$, every witness arising from $M(F, V)$ can also be obtained from $M(F', V)$ and so (5.6) holds.

Let $\mathrm{Lat}\,\mathcal{H}$ denote the Rees quotient (see Sect. 3.1) of $\mathrm{FSub}_\wedge \mathrm{Fl}\,\mathcal{H}$ by the above down set. Then $\mathrm{Lat}\,\mathcal{H} = \mathrm{Im}\,\theta \cup \{B\}$ has a natural lattice structure (see Proposition 3.1.1).

On the other hand, adding a (new) bottom element B to LR $\mathcal{H}$, we get a quasi-ordered set $\text{LR}_0\ \mathcal{H} = \text{LR}\ \mathcal{H}\ \cup\{B\}$ and we can extend θ to an onto map $\theta_0 : \text{LR}_0\ \mathcal{H} \to \text{Lat}\ \mathcal{H}$ by setting $B\theta_0 = B$.

Proposition 5.5.4. *Let $\mathcal{H}$ be a simple simplicial complex and let $R, S \in \text{LR}_0\ \mathcal{H}$. Then*

$$R \leq S \quad \text{if and only if} \quad R\theta_0 \leq S\theta_0.$$

Proof. If $R, S \neq B$, the claim follows from Proposition 5.5.1. The remaining cases are trivial. $\square$

Let ρ_0 be the equivalence in $\text{LR}_0\ \mathcal{H}$ defined by $\rho_0 = (\leq \cap \geq)$. Clearly, two lattice representations $(L, V), (L', V)$ are ρ_0-equivalent if and only if there exists some lattice isomorphism $\varphi : L \to L'$ which is the identity on V. Then the quotient $\text{LR}_0\ \mathcal{H}\ /\rho_0$ becomes a poset and by Proposition 5.5.4 the induced mapping $\overline{\theta_0} : \text{LR}_0\ \mathcal{H}\ /\rho_0 \to \text{Lat}\ \mathcal{H}$ is a poset isomorphism. Since we have already remarked that $\text{Lat}\ \mathcal{H}$ is a lattice (with the determined join), we have proved the following theorem:

Theorem 5.5.5. *Let $\mathcal{H}$ be a boolean representable simple simplicial complex. Then $\overline{\theta_0} : \text{LR}_0\ \mathcal{H}\ /\rho_0 \to \text{Lat}\ \mathcal{H}$ is a lattice isomorphism.*

The atoms of $\text{LR}_0\ \mathcal{H}$ determine the *minimal* lattice representations of $\mathcal{H}$, and the sji elements of $\text{LR}_0\ \mathcal{H}$ determine the sji lattice representations. Clearly, meet is given by intersection in $\text{Lat}\ \mathcal{H}$, collapsing into the bottom B if it does not correspond anymore to a representation of $\mathcal{H}$. But how is the determined join characterized in this lattice?

Proposition 5.5.6. *Let $\mathcal{H} = (V, H)$ be a boolean representable simple simplicial complex. Let $F, F' \in \text{Lat}\ \mathcal{H}$. Then:*

(i) $(F \vee F') = F \cup F' \cup \{Z \cap Z' \mid Z \in F,\ Z' \in F'\}$;
(ii) *If $(L, V)\theta = F$, $(L', V)\theta = F'$ and $(L'', V)\theta = (F \vee F')$, then $M(L'', V)$ is the closure of $M(L, V) \oplus_b M(L', V)$ under row sum in $\mathbb{B}^{|V|}$.*

Proof. (i) Clearly, the right hand side is the (full) $\cap$-subsemilattice of $\text{Fl}\ \mathcal{H}$ generated by $F \cup F'$.

(ii) Recall the isomorphism from Proposition 3.5.2(ii). The rows r_Z of $M(L, V)$ (respectively $M(L', V)$, $M(L'', V)$) are determined then by the flats Z in F (respectively F', $F \vee F'$). It is immediate that $r_{Z \cap Z'} = r_Z + r_{Z'}$ in $\mathbb{B}^{|V|}$, hence $M(L'', V)$ must be, up to permutation of rows, the stacking of $M(L, V)$ and $M(L', V)$, to which we add (if needed) rows which are the sum in $\mathbb{B}^{|V|}$ of rows in $M(L', V)$ and $M(L'', V)$. $\square$

Next we introduce the notion of *boolean sum* in $\text{LR}\ \mathcal{H}$. Given $(L, V), (L', V) \in \text{LR}\ \mathcal{H}$, let $(L, V)\oplus_b (L', V)$ denote the $\vee$-subsemilattice of the direct product $L \times L'$ $\vee$-generated by the diagonal

$$\Delta_V = \{(p, p) \mid p \in V\} \subseteq L \times L'.$$

Taking the determined meet, and identifying Δ_V with V as usual, it follows that $(L, V) \oplus_b (L', V) \in$ FLg. In fact, since the projection $(L, V) \oplus_b (L', V) \to (L, V)$ is a $\vee$-map which is the identity on V, it follows easily that $(L, V) \oplus_b (L', V) \in$ LR $\mathcal{H}$. But we can prove more:

Proposition 5.5.7. *Let $\mathcal{H} = (V, H)$ be a simple simplicial complex and let (L, V), $(L', V) \in$ LR $\mathcal{H}$. Then:*

(i) $(L, V)\rho \vee (L', V)\rho = ((L, V) \oplus_b (L', V))\rho$ *holds in* $\mathrm{LR}_0\,\mathcal{H} / \rho$;
(ii) $M((L, V)\rho \vee (L', V)\rho)$ *is the closure of the stacking matrix* $M(L, V) \oplus_b M(L', V)$ *under row sum in* $\mathbb{B}^{|V|}$.

Proof. (i) By the preceding comment, we have $(L, V) \leq (L, V) \oplus_b (L', V)$ and also $(L', V) \leq (L, V) \oplus_b (L', V)$, hence

$$(L, V)\rho \vee (L', V)\rho \leq ((L, V) \oplus_b (L', V))\rho.$$

Now let $(L'', V) \in$ LR $\mathcal{H}$ and suppose that $(L, V), (L', V) \leq (L'', V)$. We must show that also $(L, V) \oplus_b (L', V) \leq (L'', V)$.

Indeed, there exist $\vee$-maps $\varphi : L'' \to L$ and $\varphi' : L'' \to L'$ which fix V. Let $\varphi'' : L'' \to L \times L'$ be defined by $x\varphi'' = (x\varphi, x\varphi')$. It is easy to check that φ'' is a $\vee$-map which fixes the elements of V. Since V $\vee$-generates L'', it follows that $\mathrm{Im}\,\varphi'' \subseteq (L, V) \oplus_b (L', V)$ and we may view φ'' as a $\vee$-map from L'' to $(L, V) \oplus_b (L', V)$. Hence $(L, V) \oplus_b (L', V) \leq (L'', V)$ and (i) holds.
(ii) By Proposition 5.5.6(ii). $\square$

Since every element of a lattice is a join of sji elements, we can now state the following straightforward consequence:

Corollary 5.5.8. *Let $\mathcal{H} = (V, H)$ be a boolean representable simple simplicial complex. Then:*

(i) *Every lattice representation of $\mathcal{H}$ can be decomposed as a boolean sum (equivalently, stacking matrices and closing under row sum) of sji representations;*
(ii) *This decomposition is not unique in general, but becomes so if we choose a maximal decomposition by taking all the sji representations below.*

Examples shall be provided in Sect. 5.7.

Remark 5.5.9. Given a simple simplicial complex $\mathcal{H} = (V, H)$ and (L, V), $(L', V) \in$ LR $\mathcal{H}$, it is reasonable to identify (L, V) and (L', V) if some bijection of V induces an isomorphism $L \to L'$, and list only up to this identification in examples. However, for purposes of boolean sum decompositions, the bijection on V must be the identity.

Thus we shall devote particular attention to minimal/sji boolean representations of $\mathcal{H}$. How do these concepts relate to the flats in $\mathrm{FSub}_\wedge\mathrm{Fl}\,\mathcal{H}$ and to the matrices representing them? We start by a general remark.

Proposition 5.5.10. *Let $(L, A) \in \text{FLg}$ and let $M = M(L, A) = (m_{xa})$. Then:*

$$\text{smi}(\text{Fl}(L, A)) = \{Z_x \mid x \in \text{smi}(L)\}.$$

Proof. Indeed, since meet in $\text{Fl}(L, A)$ is intersection, the smi elements are precisely those which cannot be expressed as intersections of flats. In view of Proposition 3.5.2 (and particularly (3.9)), these are precisely the flats of the form Z_x for $x \in \text{smi}(L)$. $\square$

If we transport these notions into $M(L, V)$, then $\text{smi}(\text{Fl}(L, V))$ corresponds to the submatrix $\widehat{M}(L, V)$ determined by the nonzero rows which are not sums of other rows in $\mathbb{B}^{|V|}$. By Proposition 5.2.4(ii), $\widehat{M}(L, V)$ is still a boolean representation of $\mathcal{H}$. We have just proved that:

Corollary 5.5.11. *Let $\mathcal{H} = (V, H)$ be a simple simplicial complex and let $(L, V) \in \text{LR}\,\mathcal{H}$. Then $\widehat{M}(L, V) = M(\text{smi}(\text{Fl}(L, V)), V)$ is a boolean representation of $\mathcal{H}$.*

Note that, if we consider $\mathbb{B}$ ordered by $0 < 1$ and the product partial order in $\mathbb{B}^{|V|}$, then the rows in $\widehat{M}(L, V)$ are precisely the sji rows of $M(L, V)$ for this partial order.

Since every onto $\vee$-morphism of lattices is necessarily a $\vee$-map, we shall follow [44, Section 5.2] and call an onto $\vee$-map a $\vee$-*surmorphism*. We say that a $\vee$-surmorphism $\varphi : L \to L'$ is a *maximal proper $\vee$-surmorphism* (MPS) of lattices if $\text{Ker}\,\varphi$ is a minimal nontrivial $\vee$-congruence on L. This amounts to saying that φ cannot be factorized as the composition of two proper $\vee$-surmorphisms.

Finally, for (distinct) $a, b \in L$, we denote by $\rho_{a,b}$ the equivalence relation on L defined by

$$x\rho_{a,b} = \begin{cases} \{a, b\} & \text{if } x = a \text{ or } x = b \\ \{x\} & \text{otherwise} \end{cases}$$

Proposition 5.5.12. *Let $\mathcal{H} = (V, H)$ be a simple simplicial complex and let $(L, V), (L', V) \in \text{LR}\,\mathcal{H}$. Then the following conditions are equivalent:*

(i) $(L, V)\rho$ covers $(L', V)\rho$ in $\text{LR}_0\,\mathcal{H}$;
(ii) There exists an MPS $\varphi : L \to L'$ fixing the elements of V;
(iii) $\text{Fl}(L', V) = \text{Fl}(L, V) \setminus \{Z_b\}$ for some $b \in \text{smi}(L) \setminus \{B\}$.

Proof. (i) $\Rightarrow$ (ii). If $(L, V)\rho$ covers $(L', V)\rho$ in $\text{LR}_0\,\mathcal{H}$, then the (onto) $\vee$-map $\varphi : L \to L'$ cannot be factorized as the composition of two proper (onto) $\vee$-maps, and so φ is an MPS.

(ii) $\Rightarrow$ (iii). By Proposition A.3.1, $\text{Ker}\,\varphi = \rho_{a,b}$ for some $a, b \in L$ such that a covers b and b is smi. Therefore we may assume that $L' = L/\rho_{a,b}$. Since $(L, V) \geq (L', V)$, we have $\text{Fl}(L', V) \subseteq \text{Fl}(L, V)$ by Proposition 5.5.1. Clearly, $Z_{a \vee b} = Z_a$ and so (5.3) yields $\text{Fl}(L', V) = \text{Fl}(L, V) \setminus \{Z_b\}$.

Finally, suppose that $b = B$. Since V $\vee$-generates L, we get $a \in V$ and so $a = B$ in L', contradicting $(L', V) \in \text{LR } \mathcal{H} \subseteq \text{FLg}$ (which implies $V \subseteq L' \setminus \{B\}$). Therefore $b \neq B$ and (iii) holds.

(iii) $\Rightarrow$ (i). By Proposition 5.5.1, there is a $\vee$-map $\varphi : L \to L'$ fixing the elements of V. It follows easily from (5.3) that $\text{Ker } \varphi$ has one class with two elements and all the others are singular, hence $|L'| = |L| - 1$ and so $(L, V)\rho$ covers $(L', V)\rho$ in $\text{LR}_0 \, \mathcal{H}$. $\square$

This will help us to characterize the minimal lattice representations of $\mathcal{H}$ in terms of their flats:

Proposition 5.5.13. *Let* $\mathcal{H} = (V, H)$ *be a simple simplicial complex and let* $(L, V) \in \text{LR } \mathcal{H}$. *Then the following conditions are equivalent:*

(i) (L, V) *is minimal;*

(ii) *For every MPS* $\varphi : L \to L'$ *fixing the elements of* V, $(L', V) \notin \text{LR } \mathcal{H}$;

(iii) *For every* $b \in \text{smi}(L) \setminus \{B\}$, *the matrix obtained by removing the row* b *from* $M(L, V)$ *is not a matrix boolean representation of* $\mathcal{H}$;

(iv) *For every* $b \in \text{smi}(L) \setminus \{B\}$, $\text{Fl}(L, V) \setminus \{Z_b\} \notin \text{Im } \theta$.

Proof. (i) $\Leftrightarrow$ (ii) $\Leftrightarrow$ (iv). By Proposition 5.5.12.

(i) $\Rightarrow$ (iii). Let $b \in \text{smi}(L) \setminus \{B\}$ and let a be the unique element of L covering b. By Proposition 5.5.12, $L' = L/\rho_{a,b}$ is a lattice and $M(L', V)$ is precisely the matrix obtained by removing the bth row from $M(L, V)$. If $M(L', V)$ is a boolean representation of $\mathcal{H}$, then $(L, V)\rho$ covers $(L', V)\rho$ in $\text{LR}_0 \, \mathcal{H}$ and so (L, V) is not minimal.

(iii) $\Rightarrow$ (iv). Suppose that $\text{Fl}(L, V) \setminus \{Z_b\} \in \text{Im } \theta$ for some $b \in \text{smi}(L) \setminus \{B\}$. Then $\text{Fl}(L, V) \setminus \{Z_b\} = \text{Fl}(L', V)$ for some $(L', V) \in \text{LR } \mathcal{H}$. It is straightforward to check that $M(L', V)$ is the matrix obtained by removing the bth row from $M(L, V)$. Thus (iii) fails. $\square$

We get a similar result for the sji lattice representations:

Proposition 5.5.14. *Let* $\mathcal{H} = (V, H)$ *be a simple simplicial complex and let* $(L, V) \in \text{LR } \mathcal{H}$. *Then the following conditions are equivalent:*

(i) (L, V) *is sji;*

(ii) *Up to isomorphism, there is at most one MPS* $\varphi : L \to L'$ *fixing the elements of* V *and such that* $(L', V) \in \text{LR } \mathcal{H}$;

(iii) *There exists at most one* $b \in \text{smi}(L) \setminus \{B\}$ *such that the matrix obtained by removing the* bth *row from* $M(L, V)$ *is still a matrix boolean representation of* $\mathcal{H}$;

(iv) *There exists at most one* $b \in \text{smi}(L) \setminus \{B\}$ *such that* $\text{Fl}(L, V) \setminus \{Z_b\} \in \text{Im } \theta$.

Proof. Clearly, (L, V) is sji if and only if $(L, V)\rho$ covers exactly one element in $\text{LR}_0 \, \mathcal{H}$. Now we apply Proposition 5.5.12, proceeding analogously to the proof of Proposition 5.5.13. $\square$

5.6 Minimum Degree

Given a boolean representable simplicial complex $\mathcal{H}$, the computation of mindeg $\mathcal{H}$ constitutes naturally a major issue. An interesting question is whether mindeg $\mathcal{H}$ is achieved on minimal (sji) lattice representations.

We start by considering the following minimality concept. We call a reduced boolean representation M of $\mathcal{H}$ *rowmin* if any matrix obtained by removing a row of M is no longer a boolean representation of $\mathcal{H}$.

Proposition 5.6.1. *Let* $\mathcal{H} = (V, H)$ *be a simple simplicial complex and let* $(L, V) \in \mathrm{LR}\,\mathcal{H}$ *be minimal. Then* $\widehat{M}(L, V)$ *is rowmin.*

Proof. By Proposition 5.5.13, we cannot remove from $\widehat{M}(L, V)$ a row corresponding to some $b \in \mathrm{smi}(L) \setminus \{B\}$. Suppose now that B is smi. Then B is covered in L by a unique element a, necessarily in V since L is $\vee$-generated by V and so the unique 1 in the ath column of $\widehat{M}(L, V)$ occurs at the Bth row. Since the ath column is independent due to $\mathcal{H}$ being simple, it follows that the Bth row cannot be removed either. $\square$

However, we shall see in Sect. 5.7.1 that the converse is far from true: there may exist boolean representations of minimum degree which do not arise from minimal lattice representations.

To help us to approximate mindeg $\mathcal{H}$, we introduce the following notation for a lattice L:

$$\tilde{L} = L \setminus (\{B, T\} \cup \mathrm{at}(L)),$$

$$L\alpha = |\mathrm{smi}(L)|, \quad L\beta = |\tilde{L}|.$$

Lemma 5.6.2. *Let* $(L, V), (L', V') \in \mathrm{FLg}$ *and let* $\varphi : L \to L'$ *be a* $\vee$-*surmorphism. Then* $L\beta \geq L'\beta$.

Proof. Let $x' \in \tilde{L'}$. Since φ is onto, we have $x' = x\varphi$ for some $x \in L$. We may assume x to be maximal. Since φ is onto, it follows that $x \neq B, T$. Suppose that $x \in \mathrm{at}(L)$. Since $x' \notin \mathrm{at}(L')$, we have $x' > y' > B$ for some $y' \in L'$. Write $y' = y\varphi$. Then

$$(x \vee y)\varphi = (x\varphi \vee y\varphi) = (x' \vee y') = x'$$

and so $y \leq x$ by maximality of x. Since $x \in \mathrm{at}(L)$, we get $y = x$ or $y = B$, contradicting $x' > y' > B$. Hence $x \in \tilde{L}$. Since $x' \mapsto x$ defines an injective mapping from $\tilde{L'}$ to $\tilde{L}$, we get $L\beta \geq L'\beta$. $\square$

Given a boolean representable simplicial complex $\mathcal{H} = (V, H)$, we define

$$\mathcal{H}\alpha = \min\{L\alpha \mid (L, V) \in \mathrm{LR}\ \mathcal{H}\ \text{is minimal}\},$$
$$\mathcal{H}\beta = \min\{L\beta \mid (L, V) \in \mathrm{LR}\ \mathcal{H}\ \text{is minimal}\}.$$

Proposition 5.6.3. *Let $\mathcal{H}$ be a boolean representable simplicial complex. Then:*

(i) mindeg $\mathcal{H} \leq \mathcal{H}\alpha$;
(ii) If $\mathcal{H}$ is simple of dimension ≤ 2, then mindeg $\mathcal{H} \geq \mathcal{H}\beta$.

Proof. (i) Let $\mathcal{H} = (V, H)$. Assume that $\mathcal{H}\ \alpha = L\alpha$ for some minimal $(L, V) \in \mathrm{LR}\ \mathcal{H}$. Let $M = M(L, V)$. Clearly, we may remove from the matrix the row of zeroes corresponding to T. On the other hand, the row of $x \wedge y$ is the sum of the rows of x and y in $\mathbb{B}^{|V|}$, so successive application of Proposition 5.2.4(ii) implies that $M[\mathrm{smi}(L), V]$ is still a boolean representation of $\mathcal{H}$. Hence mindeg $\mathcal{H} \leq |\mathrm{smi}(L)| = L\alpha = \mathcal{H}\alpha$.

(ii) Assume that mindeg $\mathcal{H} = q$ and is realized by some $\{1,\dots,q\} \times V$ boolean matrix representation $M = (m_{ip})$ of $\mathcal{H}$. By Proposition 4.2.2(ii) and Lemma 5.2.1, the lattice $L = \mathrm{Fl}\ M$ is contained in $\mathrm{Fl}\ \mathcal{H}$. We claim that $|\tilde{L}| \leq q$.

Indeed, let $x \in \tilde{L}$. For $i \in \{1,\dots,q\}$, we have $Z_i = \{p \in V \mid m_{ip} = 0\}$. Since $x \neq T$, we have $x = Z_{i_1} \cap \dots \cap Z_{i_n}$ for some (distinct) $i_1, \dots, i_n \in \{1,\dots,q\}$. Since $x \notin \{B\} \cup \mathrm{at}(L)$, there exists some $y \in L$ such that $x > y > B$. Hence

$$T > Z_{i_1} \geq x > y > B \tag{5.7}$$

is a chain in $\mathrm{Fl}\ M$, hence also in $\mathrm{Fl}\ \mathcal{H}$. Since $\mathrm{ht}\,\mathrm{Fl}\ \mathcal{H} \leq 3$ by Corollary 5.2.8, (5.7) yields $x = Z_{i_1}$ and so each $x \in \tilde{L}$ determines a row of M. It follows that $|\tilde{L}| \leq q$ as claimed.

Now by Proposition 5.4.3 we may view (L, V) as a lattice representation of $\mathcal{H}$, which does not need to be minimal. However, we can always find a minimal one, say (L', V), such that $(L, V) \geq (L', V)$. In particular, there exists a $\vee$-surmorphism $\varphi : L \to L'$ and it follows from Lemma 5.6.2 that

$$\mathcal{H}\beta \leq L'\beta \leq L\beta = |\tilde{L}| \leq q = \mathrm{mindeg}\ \mathcal{H}\ .$$

$\square$

Immediately, we obtain:

Corollary 5.6.4. *Let $\mathcal{H}$ be a boolean representable simple simplicial complex of dimension ≤ 2.*

(i) If $\mathcal{H}\alpha = \mathcal{H}\beta$, then mindeg $\mathcal{H} = \mathcal{H}\alpha = \mathcal{H}\beta$.
(ii) If $\deg M = \mathcal{H}\beta$ for some boolean representation M of $\mathcal{H}$, then mindeg $\mathcal{H} = \mathcal{H}\beta$.

An example of application of this Corollary is given in Proposition 5.7.12.

Next, we make some remarks on the connections with dimension, paying special attention to the case of uniform matroids.

Lemma 5.6.5. *Let* $\mathcal{H} = (V, H)$ *be a boolean representable simple simplicial complex of dimension* d. *Then*

$$max\{d + 1, \log_2 |V|\} \leq mindeg\ \mathcal{H} \leq |H \setminus fct\ \mathcal{H}\ | \leq \sum_{i=0}^{d} \binom{|V|}{i}.$$

Proof. Let $r = $ mindeg $\mathcal{H}$ and let M be a boolean representation of minimum degree of $\mathcal{H}$. Clearly, since dim $\mathcal{H} = d$, we need at least $d + 1$ rows in M in order to have $d + 1$ independent columns. Hence $r \geq d + 1$.

On the other hand, we have 2^r possible column vectors. Since $\mathcal{H}$ is simple, all the column vectors of M must be distinct, hence $|V| \leq 2^r$ and so $r \geq \log_2 |V|$.

Now, since a row of zeroes has no place in M, we have Fl $M \subseteq$ Fl $\mathcal{H} \setminus \{V\}$ and Propositions 4.2.3 and 4.2.4 yield $r \leq |H \setminus$ fct $\mathcal{H}\ |$. Finally, since every face of dimension d of H is necessarily a facet, we get $|H \setminus$ fct $\mathcal{H}\ | \leq \sum_{i=0}^{d} \binom{|V|}{i}$. $\square$

Given $m, n \in \mathbb{N}$ with $m \leq n$, $U_{m,n} = (V, H)$ denotes the uniform simplicial complex (a matroid, actually) such that $|V| = n$ and $H = P_{\leq m}(V)$. We shall assume that $V = \{1, \ldots, n\}$. It follows from Propositions 4.2.2(iii) and 4.2.3 that

$$Fl\ U_{m,n} = P_{\leq m-1}(V) \cup \{V\}.$$

In order to illustrate the ideas of this section, we perform some calculations on the minimum degree of uniform matroids.

Theorem 5.6.6. *Let* $1 \leq m < n$. *Then*

$$\frac{1}{m}\binom{n}{m-1} \leq mindeg\ U_{m,n} \leq \binom{n-1}{m-1}.$$

Proof. Suppose that $P = (p_{ij}) \in M_{r \times n}(\mathbb{B})$ is a boolean representation of $U_{m,n}$ of degree $r = $ mindeg $U_{m,n}$. By Lemma 2.2.4, for every $X \in P_m(V) \subset H$ there exists some $Y \in$ Fl $M \cap P_{m-1}(V)$ such that $Y \subset X$. Now $|P_m(V)| = \binom{n}{m}$ and each $Y \in P_{m-1}(V)$ is contained in $n - (m-1)$ elements of $P_m(V)$. It follows that

$$r \geq \tfrac{1}{n-(m-1)}\binom{n}{m} = \tfrac{1}{n-(m-1)}\binom{n}{m-1}\tfrac{n-(m-1)}{m}$$

$$= \tfrac{1}{m}\binom{n}{m-1}.$$

For the second inequality, let $F = P_{m-1}(\{1, \ldots, n-1\})$. Write $M = $ Mat $\mathcal{H}$ and $N = M[F, V]$. Since $|F| = \binom{n-1}{m-1}$, it suffices to show that N is a boolean representation of $U_{m,n}$.

Let $X \subseteq V$ be c-independent with respect to N. Since each row of N has precisely $m - 1$ zeroes, it follows from Lemma 2.2.4 that $X \in P_{\leq m}(V)$.

Conversely, we must show that every $X \in P_{\leq m}(V)$ is c-independent with respect to N. We may assume without loss of generality that $|X| = m$ and write $X = x_1 \dots x_m$ with $x_2, \dots, x_m \neq n$. For $i = 2, \dots, m$, let $y_i = x_1$ if $x_1 \neq n$; if $x_1 = n$, choose $y_i \in V \setminus X$.

Write $X_1 = x_2 \dots x_m$. For $i = 2, \dots, m$, let $X_i = (X \setminus \{x_1, x_i\}) \cup \{y_i\}$. It is immediate that $X_i \in F$ for every $i \in \{1, \dots, m\}$. Moreover, after permuting rows and columns according to the given enumerations, we have

$$N[X_1, \dots, X_m; x_1, \dots, x_m] = \begin{pmatrix} 1 & 0 & 0 & \dots & 0 \\ ? & 1 & 0 & \dots & 0 \\ ? & 0 & 1 & \dots & 0 \\ \vdots & \vdots & \vdots & \ddots & \vdots \\ ? & 0 & 0 & \dots & 1 \end{pmatrix}$$

Thus is c-independent with respect to N. The converse implication follows from N being a submatrix of Mat $\mathcal{H}$ and Theorem 5.2.5. $\square$

We can now get a result on the asymptotics of mindeg $U_{m,n}$. Given $k \in \mathbb{N}$, the complexity class $\Theta(n^k)$ consists of all functions $\varphi : \mathbb{N} \to \mathbb{N}$ such that

$$\exists K, K' > 0 \; \exists n_0 \in \mathbb{N} \; \forall n \geq n_0 \quad K n^k \leq n\varphi \leq K' n^k.$$

Corollary 5.6.7. *Let $m \geq 1$ be fixed and consider* mindeg $U_{m,n}$ *as a function of n for $n > m$. Then* mindeg $U_{m,n} \in \Theta(n^{m-1})$.

Proof. By Theorem 5.6.6, we have

$$\text{mindeg } U_{m,n} \geq \frac{1}{m}\binom{n}{m-1} = \frac{n!}{m!(n-(m-1))!}$$
$$= \frac{n(n-1)\dots(n-(m-2))}{m!}.$$

For large enough n, we have $\frac{n}{n-(m-2)} \leq 2^{\frac{1}{m-1}}$ and so

$$\text{mindeg } U_{m,n} \geq \left(\frac{n}{2^{\frac{1}{m-1}}}\right)^{m-1} \frac{1}{m!} = \frac{n^{m-1}}{2m!}.$$

Theorem 5.6.6 yields also

$$\text{mindeg } U_{m,n} \leq \binom{n-1}{m-1} = \frac{n(n-1)\dots(n-(m-1))}{(m-1)!} \leq \frac{n^{m-1}}{(m-1)!}.$$

Therefore mindeg $U_{m,n} \in \Theta(n^{m-1})$. $\square$

5.7   Examples

We present now some examples where we succeed in identifying all the minimal and sji lattice representations, as well as computing mindeg.

5.7.1   The Tetrahedron Complexes T_3 and T_2

Write $T_3 = (V, H)$ for $V = \{1, \ldots, 4\}$ and $H = P_{\leq 3}(V) \setminus \{123\}$.

It is routine to compute $\mathrm{Fl}\, T_3 = P_{\leq 1}(V) \cup \{14, 24, 34, 123, 1234\}$. Which $F \in \mathrm{FSub}_\wedge \mathrm{Fl}\, T_3$ correspond to lattice representations (i.e. $F \in \mathrm{Im}\,\theta$)? We have the following lemma.

Lemma 5.7.1. *The following conditions are equivalent for $F \in \mathrm{FSub}_\wedge \mathrm{Fl}\, T_3$:*

(i) $F \in \mathrm{im}\,\theta$;
(ii) One of the following conditions is satisfied:

$$123 \in F \text{ and } |\{1, 2, 3\} \cap F| \geq 2, \tag{5.8}$$

$$|\{14, 24, 34\} \cap F| \geq 2. \tag{5.9}$$

Proof. (i) $\Rightarrow$ (ii). Assume that $F \in \mathrm{Im}\,\theta$ and $|\{14, 24, 34\} \cap F| \leq 1$. Without loss of generality, we may assume that $24, 34 \notin F$. Since $234 \in H$, it follows from Corollary 5.5.3 that there exists an enumeration x, y, z of 234 such that

$$\mathrm{Cl}_F(xyz) \supset \mathrm{Cl}_F(yz) \supset \mathrm{Cl}_F(z).$$

The only possibility for $\mathrm{Cl}_F(yz)$ in F is now 123. Hence $\mathrm{Cl}_F(z) \in \{2, 3\}$. By symmetry, we may assume that $2 \in F$. On the other hand, since $13 \in H$, there exists an enumeration a, b of 13 such that

$$\mathrm{Cl}_F(ab) \supset \mathrm{Cl}_F(b) = 123.$$

The only possibilities for $\mathrm{Cl}_F(b)$ in F are now 1 or 3, hence (5.8) holds.

(ii) $\Rightarrow$ (i). In view of Corollary 5.5.3, it is easy to see that any of the conditions implies $F \in \mathrm{Im}\,\theta$ (note that $4 \in F$ in the case (5.9) since it is the intersection of two 2-sets). $\square$

We consider now the minimal case.

Proposition 5.7.2. *The following conditions are equivalent for $F \in \text{FSub}_\wedge \text{Fl } T_3$:*

(i) $F = (L, V)\theta$ *for some minimal* $(L, V) \in \text{LR } T_3$;
(ii) $F = \{V, 123, i, j, \emptyset\}$ *or* $F = \{V, i4, j4, 4, \emptyset\}$ *for some distinct* $i, j \in \{1, 2, 3\}$.

Proof. By Proposition 5.5.13, (i) holds if and only if removal of some $Z \in \text{smi}(F) \setminus \{\emptyset\}$ takes us outside $\text{Im } \theta$. It follows easily from Lemma 5.7.1 that this corresponds to condition (ii). $\square$

The F in Proposition 5.7.2 lead to the lattices

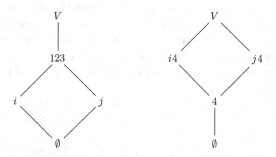

Note that, if we wish to identify the $\vee$-generating set V in these lattices, we only have to look for $\text{Cl}_F\, p$ for each $p \in V$. For instance, in the first lattice, the top element corresponds to the generator 4, and (permuting the first three columns if necessary) we get the matrix representation

$$\begin{pmatrix} 0 & 0 & 0 & 0 \\ 0 & 0 & 0 & 1 \\ 0 & 1 & 1 & 1 \\ 1 & 0 & 1 & 1 \\ 1 & 1 & 1 & 1 \end{pmatrix}$$

Following Remark 5.5.9, we can count the number of minimal lattice representations,

- Up to identity in the $\vee$-generating set V: $3 + 3 = 6$;
- Up to some bijection of V inducing a lattice isomorphism: $1 + 1 = 2$.

We remark also that the second lattice, being non atomistic, is not geometric, hence matroids can be represented by non geometric lattices!
We identify next the sji representations.

Proposition 5.7.3. *The following conditions are equivalent for $F \in \mathrm{FSub}_\wedge \mathrm{Fl}\, T_3$:*

(i) $F = (L, V)\theta$ *for some non-minimal sji* $(L, V) \in \mathrm{LR}\, T_3$;
(ii) F *is*

$$\{V, 123, i, j, 4, \emptyset\} \ \ or \ \ \{V, 123, i4, i, j, \emptyset\} \ \ or \ \ \{V, i4, j4, k, 4, \emptyset\} \qquad (5.10)$$

for some $i, j, k \in \{1, 2, 3\}$ *with* $i \neq j$.

Proof. (i) $\Rightarrow$ (ii). It follows from Proposition 5.5.14 that $(L, V) \in \mathrm{LR}T_3$ is sji if and only if there is at most one $Z \in \mathrm{smi}(F) \setminus \{\emptyset\}$ whose removal keeps us inside $\mathrm{Im}\,\theta$. Assume that F corresponds to an sji non minimal lattice representation. Suppose first that F satisfies (5.8). If none of the 2-sets $k4$ is in F, then F must contain precisely three singletons to avoid the minimal case, and one of them must be 4 to avoid having a multiple choice in the occasion of removing one of them. This gives us the first case in (5.10).

Hence we may assume that $i4 \in F$ and so also $i = 123 \cap i4$. If $j4 \in F$ for another $j \in 123$, then also $j \in F$ and we would have the option of removing either $i4$ or $j4$. Hence F contains $V, 123, i4, i, \emptyset$, and possibly any other singletons from $\{1, 2, 3\}$. In fact, it must contain at least one in view of (5.8) but obviously not both. Thus F is of the second form in (5.10) in this case.

Finally, assume that F satisfies (5.9) but not (5.8). Assume that $i4, j4 \in F$ for some distinct $i, j, \in \{1, 2, 3\}$. Then $123 \notin F$, otherwise $i, j \in F$ and we can remove either $i4$ or $j4$. Clearly, a third pair $k4$ is forbidden, otherwise we could remove any one of the three pairs. Thus F contains $V, i4, j4, 4 = i4 \cap j4, \emptyset$ and possibly any other singletons. In fact, it must contain at least one to avoid the minimal case but obviously not two, since any of them could then be removed. Thus F is of the third form in (5.10) and we are done.

(ii) $\Rightarrow$ (i). It is immediate that the cases in (5.10) lead to (L, V) non-minimal sji, the only possible removals being respectively $4, i4$ and k. $\square$

In the third case of (5.10), we must separate the subcases $k = i$ and $k \notin \{i, j\}$. Thus the sji non minimal cases lead to the lattices

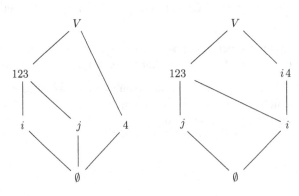

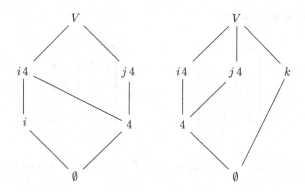

Note that the first lattice is atomistic but fails to be semimodular, but represents a matroid.

Following Remark 5.5.9 as in the minimal case, the number of sji lattice representations in both counts (which includes the minimal ones) is respectively $6 + 3 + 6 + 6 + 3 = 24$ and $1 + 1 + 1 + 1 + 1 = 5$.

It is easy to see that

$$\mathrm{Fl}\, T_3 = \{V, 123, 34, 2, 3, \emptyset\} \cup \{V, 14, 24, 1, 4, \emptyset\}$$

provides a decomposition of the top lattice representation $\mathrm{Fl}\, T_3$ as the join of two sji's. In matrix form, and in view of Proposition 5.5.7, this corresponds to decomposing the matrix

$$M(\mathrm{Fl}\, T_3, V) = \begin{pmatrix} 0 & 0 & 0 & 0 \\ 0 & 0 & 0 & 1 \\ 0 & 1 & 1 & 0 \\ 1 & 0 & 1 & 0 \\ 1 & 1 & 0 & 0 \\ 0 & 1 & 1 & 1 \\ 1 & 0 & 1 & 1 \\ 1 & 1 & 0 & 1 \\ 1 & 1 & 1 & 0 \\ 1 & 1 & 1 & 1 \end{pmatrix}$$

as the stacking of the matrices

$$
\begin{pmatrix}
0 & 0 & 0 & 0 \\
0 & 0 & 0 & 1 \\
1 & 1 & 0 & 0 \\
1 & 0 & 1 & 1 \\
1 & 1 & 0 & 1 \\
1 & 1 & 1 & 1
\end{pmatrix}
\qquad
\begin{pmatrix}
0 & 0 & 0 & 0 \\
0 & 1 & 1 & 0 \\
1 & 0 & 1 & 0 \\
0 & 1 & 1 & 1 \\
1 & 1 & 1 & 0 \\
1 & 1 & 1 & 1
\end{pmatrix}
$$

Note that the maximal decomposition of Fl T_3 as a join of sji's would include 24 factors!

We can compute also mindeg T_3:

Proposition 5.7.4. mindeg $T_3 = 3$.

Proof. Take the minimal representation defined by $F = \{V, 14, 24, 4, \emptyset\}$. By Corollary 5.5.11, we can discard the row of $M(F, V)$ corresponding to $4 = 14 \cap 24$ as well as the useless row of zeroes corresponding to V to get the matrix

$$
\widehat{M}(F, V) = \begin{pmatrix}
0 & 1 & 1 & 0 \\
1 & 0 & 1 & 0 \\
1 & 1 & 1 & 1
\end{pmatrix} \in \mathrm{LR}T_3.
$$

We cannot do better than this since H contains 3-sets. Therefore mindeg $T_3 = 3$. $\square$

The next example shows that the converse of Proposition 5.6.1 does not hold.

Example 5.7.5.

$$
M = \begin{pmatrix}
1 & 0 & 1 & 1 \\
0 & 1 & 1 & 0 \\
0 & 0 & 0 & 1
\end{pmatrix}
$$

is a boolean representation of T_3 of minimum degree (therefore rowmin) but $M \neq \widehat{M}(L, V)$ for any minimal $(L, V) \in \mathrm{LR}\, T_3$.

Indeed, by Proposition 5.7.3, $F = \{V, 123, 14, 1, 2, \emptyset\}$ defines an sji representation. Since $M = \widehat{M}(F, V)$, then M is a boolean representation of T_3 by Corollary 5.5.11. It has minimum degree by Proposition 5.7.4. However, a straightforward check of the minimal cases described in Proposition 5.7.2 shows that M does not arise from any of them.

Now write $T_2 = (V, H')$ for $H' = P_{\leq 2}(V) \cup \{123, 124\}$. It is routine to compute Fl $T_2 = P_{\leq 1}(V) \cup \{12, 1234\}$. We have the following lemma.

Lemma 5.7.6. *The following conditions are equivalent for $F \in \mathrm{FSub}_\wedge \mathrm{Fl}\, T_2$:*

(i) $F \in \operatorname{im} \theta$;
(ii) $12, i, j \in F$ for some $i \in 12$ and $j \in 34$.

Proof. (i) $\Rightarrow$ (ii). Since $123 \in H'$, it follows from Corollary 5.5.3 that there exists an enumeration x, y, z of 123 such that

$$\mathrm{Cl}_F(xyz) \supset \mathrm{Cl}_F(yz) \supset \mathrm{Cl}_F(z).$$

The only possibility for $\mathrm{Cl}_F(yz)$ in F is now 12 and so $\mathrm{Cl}_F(z) \in \{1, 2\}$. Analogously, $34 \in H'$ implies $3 \in F$ or $4 \in F$.

(ii) $\Rightarrow$ (i). Easy to check in view of Corollary 5.5.3. $\square$

We consider now the minimal case.

Proposition 5.7.7. *The following conditions are equivalent for $F \in \mathrm{FSub}_\wedge \mathrm{Fl}\, T_2$:*

(i) $F = (L, V)\theta$ for some minimal $(L, V) \in \mathrm{LR}\, T_2$;
(ii) $F = \{V, 123, i, j, \emptyset\}$ for some $i \in 12$ and $j \in 34$.

Proof. By Proposition 5.5.13, (i) holds if and only if removal of some $Z \in \mathrm{smi}(F) \setminus \{\emptyset\}$ takes us outside $\operatorname{Im} \theta$. Now we apply Lemma 5.7.6. $\square$

The F in Proposition 5.7.7 lead to the lattices

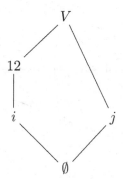

Following Remark 5.5.9, we can count the number of minimal lattice representations,

- Up to identity in the $\vee$-generating set V: 4;
- Up to some bijection of V inducing a lattice isomorphism: 1.

These are all the sji representations. Indeed, by Proposition 5.5.14, $(L, V) \in \mathrm{LR}\, T_3$ is sji if and only if there is at most one $Z \in \mathrm{smi}(F) \setminus \{\emptyset\}$ whose removal keeps us inside $\operatorname{Im} \theta$. If $1, 2 \in F$ or $3, 4 \in F$, this choice is not unique and our claim follows.

Finally, we compute $\mathrm{mindeg}\, T_2$:

Proposition 5.7.8. mindeg $T_2 = 3$.

Proof. One can check directly that

$$\begin{pmatrix} 0 & 0 & 1 & 1 \\ 0 & 1 & 1 & 1 \\ 1 & 1 & 0 & 1 \end{pmatrix}$$

is a boolean matrix representation of T_2 We cannot do better than this since H' contains 3-sets. Therefore mindeg $T_2 = 3$. $\square$

5.7.2 The Fano Matroid

Let $F_7 = (V, H)$ be the Fano matroid defined by $V = \{1, \ldots, 7\}$ and $H = P_{\le 3}(V) \setminus \{125, 137, 146, 236, 247, 345, 567\}$. We note that $\mathcal{L} = P_{\le 3}(V) \setminus H$ is precisely the set of *lines* in the Fano plane (the projective plane of order 2 over the two element field):

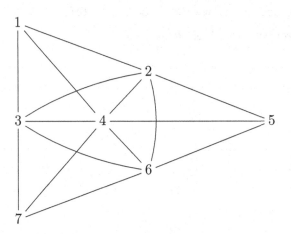

The Fano plane is an example of a partial Euclidean geometry (PEG). PEGs are studied in generality in Sect. 6.3.

Given $p \in V$, write

$$\mathcal{L}_p = \{L \in \mathcal{L} \mid p \in L\} \qquad \text{and} \qquad \mathcal{L}'_p = \{L \in \mathcal{L} \mid p \notin L\}.$$

We list a few of the properties of F_7:

(F1)   Any two distinct lines intersect at a single point.
(F2)   Every point belongs to exactly three lines.
(F3)   Any two points belong to some line.

(F4) If K consists of 5 lines, then $K \supset \mathcal{L}'_p$ for some $p \in V$.
(F5) If $p, q \in V$ are distinct, then $|\mathcal{L}'_p \cap \mathcal{L}'_q| = 2$.

Indeed, (F1)–(F3) are immediate. Then we note that (F4) follows from (F1) since the two lines not in K must intersect at some point p, hence $\mathcal{L}'_p \subset K$. Finally, it follows from (F1) and (F3) that $|\mathcal{L}_p \cap \mathcal{L}_q| = 1$, and in view of (F2) we get

$$|\mathcal{L} \setminus (\mathcal{L}'_p \cap \mathcal{L}'_q)| = |\mathcal{L}_p \cup \mathcal{L}_q| = |\mathcal{L}_p| + |\mathcal{L}_q| - |\mathcal{L}_p \cap \mathcal{L}_q| = 3 + 3 - 1 = 5,$$

thus $|\mathcal{L}'_p \cap \mathcal{L}'_q| = 7 - 5 = 2$ and (F5) holds.
    It is easy to check that

$$\mathrm{Fl}\, F_7 = P_{\leq 1}(V) \cup \mathcal{L} \cup \{V\}.$$

Indeed, the lines are obviously closed, the 2-subsets are not, and every 4-subset of V contains some facet and has therefore closure V by Proposition 4.2.3.

Lemma 5.7.9. *The following conditions are equivalent for* $F \in \mathrm{FSub}_\wedge \mathrm{Fl}\, F_7$:

(i) $F \in \mathrm{im}\,\theta$;
(ii) $\mathcal{L}'_p \subseteq F$ *for some* $p \in V$.

Proof. (i) $\Rightarrow$ (ii). Assume that $F \in \mathrm{Im}\,\theta$. If $|F \cap \mathcal{L}| \geq 5$, we are done by (F4), hence we may assume that $|F \cap \mathcal{L}| \leq 4$.
    Given $X_1, X_2, X_3 \in \mathcal{L}$ distinct, we claim that there exists some $X_4 \in \mathcal{L} \setminus \{X_1, X_2, X_3\}$ such that $\mathcal{L}'_p \not\subseteq \{X_1, X_2, X_3, X_4\}$ for every $p \in V$: indeed, there is at most one $p \in V$ such that $X_1, X_2, X_3 \in \mathcal{L}'_p$ by (F5), so it suffices to choose any $X_4 \in \mathcal{L}_p$. Since $\mathrm{Im}\,\theta$ is an up set of $\mathrm{FSub}_\wedge \mathrm{Fl}\, \mathcal{H}$ by (5.6), we may restrict ourselves to the case when $|F \cap \mathcal{L}| = 4$ and $\mathcal{L}'_p \not\subseteq F$ for every $p \in V$.
    Then $F \cap \mathcal{L}_p \neq \emptyset$ for every $p \in V$. Since $|F \cap \mathcal{L}| = 4$ implies $\sum_{i=1}^{7} |F \cap \mathcal{L}_i| = 12$, there exist distinct $p, q \in V$ such that $|F \cap \mathcal{L}_p| = |F \cap \mathcal{L}_q| = 1$. Let L (respectively L') be the unique line in $F \cap \mathcal{L}_p$ (respectively $F \cap \mathcal{L}_q$). By (F5), we have $L \neq L'$, and $|L \cup L'| = 5$ by (F1). By (F1), there is a unique $r \in V$ such that $pqr \in \mathcal{L}$. Take $s \in V \setminus (L \cup L' \cup \{r\})$. By (F1), we have $pqs \in \mathcal{H}$. It is easy to check that $\mathrm{Cl}_F(pq) = \mathrm{Cl}_F(pr) = \mathrm{Cl}_F(qr) = V$, hence by Corollary 5.5.3 we reach a contradiction. Therefore (ii) holds.
    (ii) $\Rightarrow$ (i). Let $X = xyz \in \mathcal{H}$. By (F3), we may write $x'yz, xy'z, xyz' \in \mathcal{L}$ for some $x', y', z' \in V$. We claim that

$$\{x'yz, xy'z, xyz'\} \cap \mathcal{L}'_p \neq \emptyset. \tag{5.11}$$

Indeed, if $x = p$, then $x'yz \in \mathcal{L}'_p$ and the cases $y = p$ or $z = p$ are similar. Finally, if $x, y, z \neq p$, then $x' = y' = z' = p$ would contradict (F5). Therefore (5.11) holds.

We may assume that $x'yz \in \mathcal{L}'_p \subseteq F$ and so $\mathrm{Cl}_F(yz) = x'yz$. On the other hand, in view of (F1), y belongs to precisely one line in $\mathcal{L}_p$, hence y belongs to at least two lines in F and so $y \in F$ since F is closed under intersection. Thus

$$\mathrm{Cl}_F(\emptyset) = \emptyset \subset \mathrm{Cl}_F(y) = y \subset x'yz = \mathrm{Cl}_F(yz) \subset V = \mathrm{Cl}_F(xyz)$$

and so $F \in \mathrm{Im}\,\theta$ by Corollary 5.5.3. $\square$

Proposition 5.7.10. *The following conditions are equivalent for $F \in \mathrm{FSub}_\wedge \mathrm{Fl}\, F_7$:*

(i) $F = (L, V)\theta$ for some minimal $(L, V) \in \mathrm{LR}\, F_7$;
(ii) $F = \mathcal{L}'_p \cup \{V\} \cup P_1(V \setminus \{p\})$ for some $p \in V$.

Proof. (i) $\Rightarrow$ (ii). Similar to the proof of Proposition 5.7.2, the lattice (L, V) is a minimal lattice representation if and only if removal of any $Z \in \mathrm{smi}(F) \setminus \{\emptyset\}$ takes us outside $\mathrm{Im}\,\theta$. By Lemma 5.7.9, we must have $F \cap \mathcal{L} = \mathcal{L}'_p$ for some $p \in V$. Similar to the proof of Lemma 5.7.9, we get $q \in F$ for every $q \in V \setminus \{p\}$. Since $\emptyset, V \in F$ necessarily and $p \in F$ would be removable, (ii) holds.

(ii) $\Rightarrow$ (i). If (ii) holds, then we can remove no line (by Lemma 5.7.9) and we can remove no point either since all the points in F occur as intersections of lines in F, and a $\wedge$-subsemilattice of $\mathrm{Fl}\, F_7$ must be closed under intersection. $\square$

Writing $F \cap \mathcal{L} = \{P, Q, R, S\}$ and denoting by XY the intersection of $X, Y \in F \cap \mathcal{L}$, we see that the minimal boolean representations are, up to isomorphism, given by the lattice

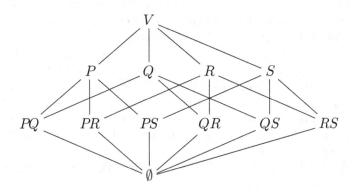

Up to congruence, the matrix representation $\widehat{M}(F, V)$ (where we may remove all the rows of $M(F, V)$ corresponding to non smi elements of F) is then of the form

$$\begin{pmatrix} 0 & 0 & 1 & 1 & 0 & 1 & 1 \\ 0 & 1 & 1 & 0 & 1 & 0 & 1 \\ 1 & 0 & 0 & 1 & 1 & 0 & 1 \\ 1 & 1 & 0 & 0 & 0 & 1 & 1 \end{pmatrix} \tag{5.12}$$

Following Remark 5.5.9 as in Sect. 5.7.1, the number of minimal lattice representations in both counts is respectively 7 and 1.

Proposition 5.7.11. *The following conditions are equivalent for* $F \in \mathrm{FSub}_\wedge \mathrm{Fl}\, F_7$:

(i) $F = (L, V)\theta$ *for some non-minimal sji* $(L, V) \in \mathrm{LR}\, F_7$;
(ii) F *satisfies one of the following conditions:*

$$F = \mathcal{L}'_p \cup \{V\} \cup P_1(V) \text{ for some } p \in V; \tag{5.13}$$

$$|F \cap \mathcal{L}| = 5 \text{ and } |F \cap P_1(V)| = 6. \tag{5.14}$$

Proof. The non-minimal sji cases are naturally divided into two categories: those which admit a (unique) removal of a 3-set, and those which admit a (unique) removal of a 1-set.

By Lemma 5.7.9, in the first category we must have $|F \cap \mathcal{L}| = 5$. By (F5), this implies $F \cap \mathcal{L} = \mathcal{L}'_p \cup \{X\}$ for some $X \in \mathcal{L}_p$. As remarked before, this implies $P_1(V \setminus \{p\}) \subseteq F$, but the point p might be removable. Thus the first category corresponds precisely to the condition (5.14).

In view of Lemma 5.7.9, in the second category we must have $F = \mathcal{L}'_p$ for some $p \in V$, and by Proposition 5.7.10 it must correspond precisely to the condition (5.13). $\square$

The lattice corresponding to (5.13) must be of the form

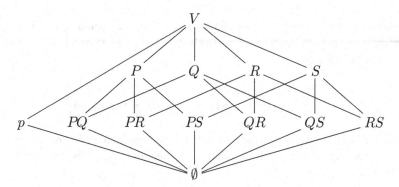

Up to congruence, the matrix representation $\widehat{M}(F, V)$ is then of the form

$$\begin{pmatrix} 0 & 0 & 1 & 1 & 0 & 1 & 1 \\ 0 & 1 & 1 & 0 & 1 & 0 & 1 \\ 1 & 0 & 0 & 1 & 1 & 0 & 1 \\ 1 & 1 & 0 & 0 & 0 & 1 & 1 \\ 1 & 1 & 1 & 1 & 1 & 1 & 0 \end{pmatrix}$$

In the case (5.14), write $\mathcal{L}'_p = \{P, Q, R, S\}$ and $F \cap \mathcal{L} = \mathcal{L}'_p \cup \{Z\}$, with $Z = pab$. Since every $x \in V \setminus \{p\}$ is intersection of lines in $\mathcal{L}'_p$, and in view of (F1), we may assume that $a = P \cap S$ and $b = Q \cap R$. We get the lattice

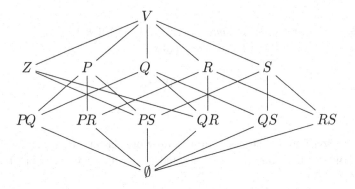

Up to congruence, the matrix representation $\widehat{M}(F, V)$ is then of the form

$$\begin{pmatrix} 0 & 0 & 1 & 1 & 0 & 1 & 1 \\ 0 & 1 & 1 & 0 & 1 & 0 & 1 \\ 1 & 0 & 0 & 1 & 1 & 0 & 1 \\ 1 & 1 & 0 & 0 & 0 & 1 & 1 \\ 1 & 1 & 1 & 1 & 0 & 0 & 0 \end{pmatrix}$$

Following Remark 5.5.9 as in the minimal case, the number of sji lattice representations in both counts is respectively $7 + 7 + 21 = 35$ and $1 + 1 + 1 = 3$.

It is easy to see that

$$\text{Fl } F_7 = \{V, 125, 146, 236, 345, 567, 1, 2, 3, 4, 5, 6, \emptyset\}$$
$$\cup \{V, 137, 146, 236, 247, 567, 1, 2, 3, 4, 6, 7, \emptyset\}$$

provides a decomposition of the top lattice representation Fl F_7 as the join of two sji's. In matrix form, and in view of Proposition 5.5.7, this corresponds to decomposing the matrix

$$\widehat{M}(\text{Fl } F_7, V) = \begin{pmatrix} 0 & 0 & 1 & 1 & 0 & 1 & 1 \\ 0 & 1 & 0 & 1 & 1 & 1 & 0 \\ 0 & 1 & 1 & 0 & 1 & 0 & 1 \\ 1 & 0 & 0 & 1 & 1 & 0 & 1 \\ 1 & 0 & 1 & 0 & 1 & 1 & 0 \\ 1 & 1 & 0 & 0 & 0 & 1 & 1 \\ 1 & 1 & 1 & 1 & 0 & 0 & 0 \end{pmatrix}$$

as

$$\begin{pmatrix} 0\ 0\ 1\ 1\ 0\ 1\ 1 \\ 0\ 1\ 1\ 0\ 1\ 0\ 1 \\ 1\ 0\ 0\ 1\ 1\ 0\ 1 \\ 1\ 1\ 0\ 0\ 0\ 1\ 1 \\ 1\ 1\ 1\ 1\ 0\ 0\ 0 \end{pmatrix} \oplus_b \begin{pmatrix} 0\ 1\ 0\ 1\ 1\ 1\ 0 \\ 0\ 1\ 1\ 0\ 1\ 0\ 1 \\ 1\ 0\ 0\ 1\ 1\ 0\ 1 \\ 1\ 0\ 1\ 0\ 1\ 1\ 0 \\ 1\ 1\ 1\ 1\ 0\ 0\ 0 \end{pmatrix}$$

where we depict only the rows corresponding to the smi elements of the lattices (minus the top).

Note that the maximal decomposition of Fl F_7 as join of sji's would include 35 factors.

Proposition 5.7.12. mindeg $F_7 = 4$.

Proof. It follows from our complete description of the minimal lattice representations that $F_7\alpha = 4 = F_7\beta$. Thus mindeg $F_7 = 4$ by Corollary 5.6.4. $\square$

Further information on the Fano plane and boolean representations, under a different perspective, can be found in [42].

5.7.3 The Uniform Matroid $U_{3,n}$

We shall analyse in this subsection the uniform matroids of the form $U_{3,n}$ with $n \geq 5$. For the simpler cases $n = 3, 4$, we can adapt the discussion of T_3 in Sect. 5.7.1.

Given $F \in \mathrm{FSub}_\wedge \mathrm{Fl}\, U_{3,n}$, we define a graph $F\gamma = (V, E)$ with $pq \in E$ whenever p, q are distinct and $pq \notin F$. This graph is actually the complement of the graph of flats of $U_{3,n}$ (see Sect. 6.4). We prove that:

Lemma 5.7.13. *Let $n \geq 5$. Then the following conditions are equivalent for $F \in \mathrm{FSub}_\wedge \mathrm{Fl}\, U_{3,n}$:*

(i) $F \in \mathrm{im}\,\theta$;
(ii) gth $F\gamma > 3$ *and* $|V \setminus F| \leq 1$.

Proof. (i) $\Rightarrow$ (ii). Suppose that gth $F\gamma = 3$. Then there exist distinct $p, q, r \in V$ such that $pq, pr, qr \notin F$. Hence $\mathrm{Cl}_F(xy) = V$ for all distinct $x, y \in \{p, q, r\}$. Since $pqr \in H$, this contradicts $F \in \mathrm{Im}\,\theta$ in view of Corollary 5.5.3. Thus gth $F\gamma > 3$.

Suppose next that $x, y \in V \setminus F$ are distinct. Let $a, b, c \in V \setminus \{x, y\}$ be distinct. By Corollary 5.5.3, xya admits an enumeration x_1, x_2, x_3 satisfying

$$\mathrm{Cl}_F(x_1x_2x_3) \supset \mathrm{Cl}_F(x_2x_3) \supset \mathrm{Cl}_F(x_3). \tag{5.15}$$

Since $F \subseteq P_{\leq 2}(V) \cup \{V\}$, we get $x_3, x_2 x_3 \in F$ and so $x_3 = a$ and $i_a a \in F$ for some $i_a \in \{x, y\}$. Similarly, $i_b b, i_c c \in F$ for some $i_b, i_c \in \{x, y\}$. Since $|\{i_a, i_b, i_c\}| \leq 2$, we may assume that $i_a = i_b = x$, hence $x = i_a a \cap i_b b \in F$, a contradiction. Therefore $|V \setminus F| \leq 1$.

(ii) $\Rightarrow$ (i). Let $x, y, z \in V$ be distinct. By Corollary 5.5.3, it suffices to show that xyz admits an enumeration x_1, x_2, x_3 satisfying (5.15). Since gth $F\gamma > 3$, we have $\{xy, xz, yz\} \cap F \neq \emptyset$. We may assume that $xy \in F$. Since $|V \setminus F| \leq 1$, we have either $x \in F$ or $y \in F$. In any case, (5.15) is satisfied by some enumeration of x, y, z and we are done. $\square$

We discuss next the minimal and sji cases.

Proposition 5.7.14. *The following conditions are equivalent for $F \in \mathrm{FSub}_\wedge \mathrm{Fl}\, U_{3,n}$:*

(i) $F = (L, V)\theta$ *for some minimal* $(L, V) \in \mathrm{LR}\, U_{3,n}$;
(ii) gth $F\gamma > 3$, diam $F\gamma = 2$ *and*

$$\mathrm{maxdeg}\, F\gamma \geq |V| - 2 \Rightarrow |V \setminus F| = 1. \tag{5.16}$$

Proof. (i) $\Rightarrow$ (ii). By Proposition 5.5.13, the minimal cases are once more characterized by the following property: removal of some $X \in \mathrm{smi}(F) \setminus \{\emptyset\}$ must make condition (ii) in Lemma 5.7.13 fail. It is easy to see that $\mathrm{smi}(F) \setminus \{\emptyset\}$ contains precisely the 2-sets and the points which are not intersections of 2-sets in F, i.e. vertices of degree $\geq |V| - 2$ in $F\gamma$.

By Lemma 5.7.13, we have gth $F\gamma > 3$ and $|V \setminus F| \leq 1$. Suppose that $\mathrm{maxdeg}\, F\gamma \geq |V| - 2$. Then there exists some $p \in V$ such that p occurs at most in one 2-set in F. Thus $p \in \mathrm{smi}(F)$. If $V \subseteq F$, it follows from Proposition 5.5.10 that condition (iv) of Proposition 5.5.13 fails if we remove p from F (since condition (ii) in Lemma 5.7.13 is still satisfied). Hence $|V \setminus F| = 1$.

Finally, since $|V| > 2$ and gth $F\gamma > 3$, we have diam $F\gamma \geq 2$. Suppose that $x, y \in V$ lie at distance >2 in $F\gamma$. Then we could add an edge $x \text{ ——— } y$ and still satisfy condition (ii) in Lemma 5.7.13. Since adding an edge corresponds to removal of the smi xy from F, this contradicts (L, V) being minimal.

(ii) $\Rightarrow$ (i). Since diam $F\gamma = 2$, it is clear that we cannot add any extra edge and keep gth $F\gamma > 3$, hence removal of 2-sets from F will takes us out of $\mathrm{Im}\,\theta$ by Lemma 5.7.13. On the other hand, also in view of Lemma 5.7.13, we can only remove a point from F if $V \subseteq F$, and by (5.16) this can only happen if $\mathrm{maxdeg}\, F\gamma < |V| - 2$. However, as remarked before, this implies that no point is in $\mathrm{smi}(F)$. Therefore (L, V) is minimal as claimed. $\square$

Proposition 5.7.15. *The following conditions are equivalent for $F \in \mathrm{FSub}_\wedge \mathrm{Fl}\, U_{3,n}$:*

(i) $F = (L, V)\theta$ for some sji $(L, V) \in \mathrm{LR}\, U_{3,n}$;
(ii) $\mathrm{gth}\, F\gamma > 3$ and one of the following cases holds:

$$\text{there exists a unique } uv \in P_2(V) \text{ such that } d(u, v) > 2 \tag{5.17}$$
$$\text{in } F\gamma \text{ and (5.16) holds;}$$

$$\mathrm{diam}\, F\gamma = 2 \quad \text{and} \quad (\, F\gamma \cong K_{2,n} \Rightarrow |V \setminus F| = 1\,). \tag{5.18}$$

Proof. (i) $\Rightarrow$ (ii). Lemma 5.7.13 yields gth $F\gamma > 3$, which implies diam $F\gamma \geq 2$.

Suppose first that diam $F\gamma = 2$. As remarked in the proof of Proposition 5.7.14, we cannot remove a 2-set from F, and smi points correspond to vertices of degree $\geq |V| - 2$ in $F\gamma$. Therefore there is at most one such vertex in $F\gamma$. Since $K_{2,n}$ has two, then (5.18) must hold.

Finally, assume that diam $F\gamma > 2$. Then there exist $u, v \in V$ at distance 3 in $F\gamma$, and adding an edge u —— v does not spoil condition (ii) of Lemma 5.7.13. Since (L, V) is sji, then the 2-set uv is unique. Similarly to the proof of Proposition 5.7.14, (5.16) must hold to prevent removal of an smi point. Therefore (5.17) holds.

(ii) $\Rightarrow$ (i). If (5.17) holds, then the unique edge that can be added to the graph and keep its girth above 3 is u —— v. On the other hand, since (5.16) holds, the possibility of removal of an smi point is excluded. Thus (L, V) is sji in this case.

Assume now that (5.18) holds. We cannot remove a 2-set from F in view of Lemma 5.7.13, since adding an edge to a graph of diameter 2 implies girth 3, forbidden in view of Lemma 5.7.13. On the other hand, having an option on removing an smi point would imply the existence of two points of degree $\geq |V| - 2$, which in view of gth $F\gamma > 3$ implies $F\gamma \cong K_{2,n}$. But in view of (5.18) and previous comments, only one of these points can be present on F. Thus (L, V) is sji also in this case. $\square$

It is now a simple exercise, for instance, to check that the minimal representations of $U_{3,6}$ correspond (up to permutation of vertices) to the graphs

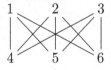 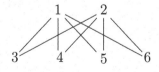

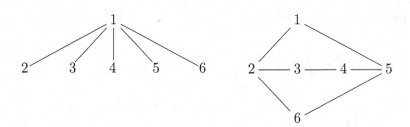

and to $F_1, F_2, F_3, F_4 \in \text{FSub}_\wedge \text{Fl } \mathcal{H}$ given respectively by

$F_1 = \{V, 12, 13, 23, 45, 46, 56, 1, 2, 3, 4, 5, 6, \emptyset\};$
$F_2 = \{V, 12, 34, 35, 36, 45, 46, 56, 1, 3, 4, 5, 6, \emptyset\};$
$F_3 = \{V, 23, 24, 25, 26, 34, 35, 36, 45, 46, 56, 2, 3, 4, 5, 6, \emptyset\};$
$F_4 = \{V, 13, 14, 16, 24, 25, 35, 36, 46, 1, 2, 3, 4, 5, 6, \emptyset\}.$

The corresponding lattices are now

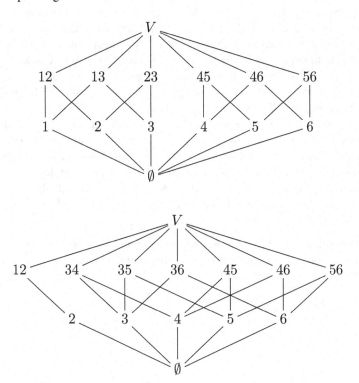

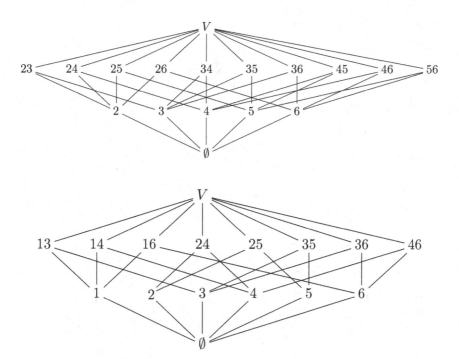

The non-minimal sji representations of $U_{3,6}$ can be easily computed. In fact, it is easy to see that if (5.17) holds, then by adding an edge u — v to the graph $F\gamma$ we get a graph of diameter 2 and still girth >3. The converse is not true, but a brief analysis of all the possible removals of one edge from a minimal case graph to reach (5.17) gives us all such sji representations.

Those of type (5.18) are obtained by adding the seventh point to the minimal representation given by $K_{1,5}$ (the other types already have the seven points or are excluded by the implication in (5.18)).

Therefore the graphs corresponding to the sji representations of type (5.17) are

obtained by removing an edge from $K_{3,3}$ and $K_{2,4}$, respectively. Adding the (essentially unique) case (5.18) representation, we obtain types

$F_5 = \{V, 12, 13, 23, 34, 45, 46, 56, 1, 2, 3, 4, 5, 6, \emptyset\};$
$F_6 = \{V, 12, 23, 34, 35, 36, 45, 46, 56, 2, 3, 4, 5, 6, \emptyset\};$
$F_7 = \{V, 23, 24, 25, 26, 34, 35, 36, 45, 46, 56, 1, 2, 3, 4, 5, 6, \emptyset\}.$

The corresponding lattices are

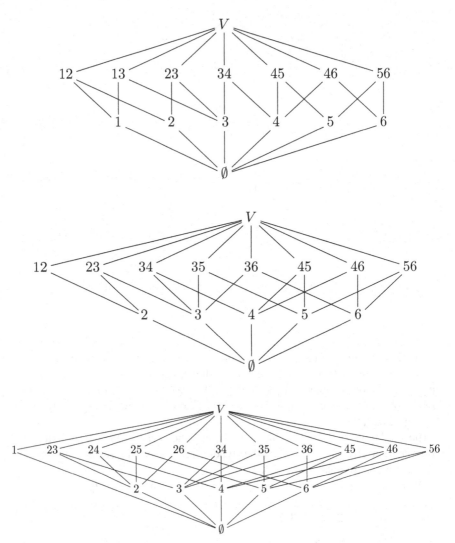

It is easy to count $20 + 30 + 6 + 180 = 236$ minimal lattice representations for $U_{3,6}$ only (but they reduce to $1 + 1 + 1 + 1 = 4$ in the alternative counting of Remark 5.5.9)! The sji's (including the minimal cases) amount to $236 + 90 + 120 + 6 = 452$ and $4 + 1 + 1 + 1 = 7$ in both countings.

Note that the lattices in the examples in which $V \subseteq F$, after removal of the top and bottom elements, are essentially the *Levi graphs* [13] of the graphs $F\gamma$. The Levi graph of $F\gamma$ can be obtained by introducing a new vertex at the midpoint of every edge (breaking thus the original edge into two), and the connection to the lattice is established by considering that each of the new vertices lies above its two adjacent neighbors.

Note also that famous graphs of girth >3 and diameter 2 such as the *Petersen graph* [16, Section 6.6] turn out to encode minimal representations via the function γ (in $U_{3,10}$, since the Petersen graph has 10 vertices).

Finally, we compute mindeg$U_{3,n}$:

Proposition 5.7.16. *For $n \geq 3$,*

$$\text{mindeg } U_{3,n} = \begin{cases} \frac{n(n-2)}{4} & \text{if } n \geq 6 \text{ and even} \\ \frac{(n-1)^2}{4} & \text{if } n \geq 6 \text{ and odd} \\ 5 & \text{if } n = 5 \\ 3 & \text{if } n = 3 \text{ or } 4 \end{cases}$$

Proof. Assume first that $n = 2m$ with $m \geq 3$. We assume that M is an $R \times V$ boolean representation of $U_{3,2m}$ with minimum degree. By Proposition 5.2.4(ii), we can add all the boolean sums of rows in M and still have a boolean representation of $U_{3,2m}$, and we can even add a row of zeroes (we are in fact building the matrix $M^\mu \in \mathcal{M}$ from Sect. 3.4). Now by Proposition 3.5.4 we have $M^\mu = M(L, V)$ for some $(L, V) \in \text{LR } U_{3,2m}$, and so $F = (L, V)\theta$ satisfies gth $F\gamma > 3$ by Lemma 5.7.13. By Turán's Theorem [16, Theorem 7.1.1], the maximum number of edges in a triangle-free graph with $2m$ vertices is reached by the complete bipartite graph $K_{m,m}$ which has m^2 edges. Therefore $F\gamma$ has at most m^2 edges. Since 2^V has $\binom{2m}{2} = m(2m-1)$ 2-sets, it follows that F has at least $m(2m-1) - m^2 = m(m-1)$ 2-sets. Since the 2-sets represent necessarily smi elements of M^μ, it follows that $M = \widehat{M}(L, V)$ has at least $m(m-1)$ elements and so mindeg $U_{3,2m} \geq m(m-1)$. Equality is now realized through $F\gamma = K_{m,m}$. Note that in this case no vertex has degree $\geq |V| - 2$, hence all the points are intersections of 2-sets in F and so the smi rows of the matrix are precisely the $k(k-1)$ rows defined by the complement graph of $K_{m,m}$. Therefore mindeg $U_{3,2m} = m(m-1)$.

Assume now that $n = 2m + 1$ with $m \geq 3$. The argument is similar to the proof of the preceding case, so we just discuss the differences. Again by Turán's Theorem [16, Theorem 7.1.1], the maximum number of edges in a triangle-free graph with $2m + 1$ vertices is reached by the complete bipartite graph $K_{m,m+1}$ which has $(m+1)m$ edges. Therefore $F\gamma$ has at most $(m+1)m$ edges. Since 2^V has $\binom{2m+1}{2} = (2m+1)m$ 2-sets, it follows that F has at least $(2m+1)m - (m+1)m = m^2$ 2-sets. Note that, since we have $m \geq 3$ no vertex has degree $\geq |V| - 2$ in $K_{m,m+1}$.

Assume now that $n = 5$. The preceding argument shows that F has at least 4 2-sets but this time in $K_{2,3}$ there is a vertex of degree 3, implying the presence of an smi point in F. Therefore the above arguments yield only $4 \leq$ mindeg $U_{3,5} \leq 5$. Suppose that there exists some $M = (m_{ij}) \in M_{4\times 5}(\mathbb{B})$ representing $U_{3,5}$. Note that in view of Lemma 5.2.1 we may assume that no row of M has more than two zeroes. Let $F = \text{Fl } M \cap P_2(V)$. In view of Corollary 5.5.3, every $X \in P_3(V)$ must contain some $Y \in F$. Now it is straightforward to check that any graph with 5 vertices and at most 4 edges admits a 3-anticlique except (up to renaming of vertices)

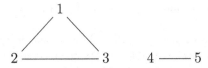

Therefore the unique possibility is to have (up to congruence) a matrix of the form

$$\begin{pmatrix} 0 & 0 & 1 & 1 & 1 \\ 0 & 1 & 0 & 1 & 1 \\ 1 & 0 & 0 & 1 & 1 \\ 1 & 1 & 1 & 0 & 0 \end{pmatrix}$$

But then 45 is not c-independent and we reach a contradiction. Therefore mindeg $U_{3,5} = 5$.

For the cases $n = 3$ and $n = 4$, it is immediate that the matrices

$$\begin{pmatrix} 1 & 0 & 0 \\ 0 & 1 & 0 \\ 0 & 0 & 1 \end{pmatrix} \quad \text{and} \quad \begin{pmatrix} 0 & 0 & 1 & 1 \\ 1 & 0 & 0 & 1 \\ 1 & 1 & 0 & 0 \end{pmatrix}$$

constitute representations of minimum degree. $\square$

5.7.4 Steiner Systems

A *Steiner system* with parameters $t < r < n$ is an ordered pair $(V, \mathcal{B})$, where $|V| = n$ and $\mathcal{B} \subseteq P_r(V)$ (the set of *blocks*) is such that each t-subset of V is contained in exactly one block. It follows that $|B \cap B'| < t$ for all distinct $B, B' \in \mathcal{B}$. For details on Steiner systems, the reader is referred to [1, Chapter 8].

We denote by $S(t, r, n)$ the class of all Steiner systems with parameters $t < r < n$. Two Steiner systems $(V, \mathcal{B})$ and $(V', \mathcal{B}')$ are said to be *congruent* if there exists a bijection $\varphi : V \to V'$ inducing a bijection on the blocks. If there exists a unique element of $S(t, r, n)$ up to congruence, it is common to use $S(t, r, n)$ to denote it.

If $(V, \mathcal{B}) \in S(t, r, n)$ and $p \in V$, let $\mathcal{B}_{(p)} = \{ X \subseteq V \setminus \{p\} \mid X \cup \{p\} \in \mathcal{B} \}$. It is easy to see that $(V \setminus \{p\}, \mathcal{B}_{(p)}) \in S(t-1, r-1, n-1)$. It is called a *derived system* of $(V, \mathcal{B})$.

A Steiner system in $S(2, 3, n)$ (respectively $S(3, 4, n)$) is called a *Steiner triple system* (respectively *Steiner quadruple system*) and the notation

$$S(2, 3, n) = STS(n), \quad S(3, 4, n) = SQS(n)$$

is standard. It is known that $SQS(n) \neq \emptyset$ if and only if $n \equiv 2 \mod 6$ or $n \equiv 4 \mod 6$. It is known that $SQS(8)$ and $SQS(10)$ are unique, while $SQS(14)$ has 4 congruence classes and $SQS(16)$ has 1,054,163. We present next a construction of $SQS(8)$. We will use this description in Theorem 5.7.18(iii) to show that the minimum degree of the paving matroid defined below corresponding to $SQS(8)$ is 6.

Consider the cube C depicted by

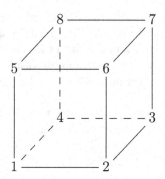

Write $V = \{1, \ldots, 8\}$ and let F denote the set of all 4-subsets of V which correspond to faces of C. We define

$$\mathcal{B} = \{X \in P_4(V) \mid |X \cap Y| \text{ is even for every } Y \in F\}.$$

It is easy to see that $\mathcal{B}$ contains 14 elements:

- The 6 faces 1234, 1458, 1256, 2367, 3478, 5678;
- The 6 diagonal planes 1278, 1357, 1467, 2358, 2468, 3456;
- The 2 twisted planes 1368, 2457.

Then $(V, \mathcal{B}) = SQS(8)$.

It is easy to check that $\mathcal{B}$ is closed under complement. Moreover,

$$|B \cap B'| \text{ is even for all } B, B' \in \mathcal{B}. \tag{5.19}$$

Indeed, if $|B \cap B'|$ is odd, we may assume that $|B \cap B'| = 3$ by replacing B by its complement if necessary. Then a 3-subset of V would be contained in two distinct blocks, a contradiction.

We note also that the Fano plane of Sect. 5.7.2 (which is $STS(7) = S(2, 3, 7)$) is a derived system of $SQS(8) = S(3, 4, 8)$.

Now every $(V, \mathcal{B}) \in S(r - 1, r, n)$ induces a paving matroid $(V, H(\mathcal{B})) \in$ BPav$(r - 1)$ (boolean representable in view of Theorem 5.2.10) defined by

$$H(\mathcal{B}) = P_{\leq r}(V) \setminus \mathcal{B}.$$

Its lattice of flats is easy to compute:

Lemma 5.7.17. *Let* $(V, \mathcal{B}) \in S(r-1, r, n)$. *Then:*

(i) $\mathrm{Fl}(V, H(\mathcal{B})) = P_{\leq r-2}(V) \cup \mathcal{B} \cup \{V\}$;
(ii) $\mathrm{smi}(\mathrm{Fl}(V, H(\mathcal{B}))) = \mathcal{B}$.

Proof. (i) We have $P_{\leq r-2}(V) \cup \{V\} \subseteq \mathrm{Fl}(V, H(\mathcal{B}))$ by Proposition 4.2.2. Let $B \in \mathcal{B}$ and suppose that $X \in H(\mathcal{B}) \cap 2^B$ and $p \in V \setminus B$. We may assume that $|X| = r - 1$. Since B is the unique block containing X, we have $X \cup \{p\} \notin \mathcal{B}$ and so $X \cup \{p\} \in H(\mathcal{B})$. Thus $B \in \mathrm{Fl}(V, H(\mathcal{B}))$.

Next consider $X \in P_{r-1}(V)$. Since $X \cup \{p\} \in \mathcal{B}$ for some (unique) $p \in V \setminus X$, it follows that $X \notin \mathrm{Fl}(V, H(\mathcal{B}))$. Finally, assume that $X \subset V$ is such that $|X| \geq r$ and $X \notin \mathcal{B}$. If every r-subset of X is in $\mathcal{B}$, then $|X| > r$ and some $(r-1)$-subset of X must be contained in two distinct blocks, a contradiction. Hence X must contain a facet and so $\mathrm{Cl}\, X = V$ by Proposition 4.2.3. Therefore $X \notin \mathrm{Fl}(V, H(\mathcal{B}))$.

(ii) It suffices to show that every $(r-2)$-subset X of V is an intersection of blocks. Indeed, for every $p \in V \setminus X$, there exists some $p' \in V$ such that $X \cup \{p, p'\} \in \mathcal{B}$. Suppose that $q \in \cap_{p \in V \setminus X} \{p, p'\}$. Since $|V \setminus X| \geq 3$, it follows that there exist two distinct $B, B' \in \mathcal{B}$ such that $B \cap B' = X \cup \{q\}$, a contradiction. Thus $X = \cap_{p \in E \setminus X} (X \cup \{p, p'\})$ as required. $\square$

Theorem 5.7.18. *Let* $(V, \mathcal{B}) \in S(r-1, r, n)$. *Then:*

(i) $M(\mathcal{B}, V)$ *is a boolean representation of* $(V, H(\mathcal{B})) \in BPav(r-1)$;
(ii) $\mathrm{mindeg}\,(V, H(\mathcal{B})) \leq |\mathcal{B}|$;
(iii) *If* $r = 4$ *and* $n = 8$, *then* $\mathrm{mindeg}\,(V, H(\mathcal{B})) = 6$.

Proof. (i) By Theorems 5.2.5 and 5.2.10, $M(\mathrm{Fl}(V, H(\mathcal{B})))$ is a boolean representation of $(V, H(\mathcal{B}))$. By Corollary 5.5.11, we only need to keep the rows corresponding to smi elements. Now the claim follows from Lemma 5.7.17(ii).

(ii) Immediate from (i).

(iii) Let F denote the set of all 4-subsets of V which correspond to faces of the cube C. We show that $M = M(F, V)$ is a boolean representation of $(V, H(\mathcal{B}))$. Since $F \subseteq \mathrm{Fl}(V, H(\mathcal{B}))$, in view of (i) and Lemma 2.2.3(i) it suffices to show that every 4-subset in $H(\mathcal{B})$ is c-independent with respect to M.

Let $X \in H(\mathcal{B})$ with $|X| = 4$ (note that, being a matroid, $(V, H(\mathcal{B}))$ is pure). By definition of $\mathcal{B}$, there exists some $Y_1 \in F$ such that $|X \cap Y_1|$ is odd. Exchanging Y_1 by its complement if needed, we may assume that $|X \cap Y_1| = 3$. Using the symmetries of the cube, we may assume without loss of generality that $Y_1 = 5678$ and $X = x567$. Let $Y_2 = 2367$, $Y_3 = 3478$ and $Y_4 = 1234$. Then $M[Y_1, Y_2, Y_3, Y_4; x, 5, 6, 7]$ is a lower unitriangular submatrix of M and so X is c-independent. Thus M is a boolean representation of $(V, H(\mathcal{B}))$ and so $\mathrm{mindeg}(V, H(\mathcal{B})) \leq 6$.

Now suppose that M is a (reduced) boolean representation of $(V, H(\mathcal{B}))$ with degree <6. By Theorem 5.2.5, we have $M = M(F, E)$ for some $F \subseteq \mathrm{Fl}(V, H(\mathcal{B}))$. Let $X \in \mathcal{B} \setminus F$. Using the symmetries of the cube, we may assume that $X = 1234$. Let $Y = 1238 \in P_4(V) \setminus \mathcal{B} \subseteq H(\mathcal{B})$.

Since Y is then c-independent with respect to M, then Y has a witness in M by Proposition 2.2.6. Hence there exists some $Z \in F$ such that $|Y \cap Z| = 3$. Thus $Z \in \mathcal{B}$ by Lemma 5.7.17(i), a contradiction since the only face of the cube sharing 3 elements with Y is the absent X. Therefore $\mathrm{mindeg}(V, H(\mathcal{B})) = 6$. $\square$

We are interested in generalizing Theorem 5.7.18(iii) (see Question 9.1.3(iv)).

Chapter 6
Paving Simplicial Complexes

We devote this chapter to the particular case of paving simplicial complexes, with special emphasis on the case of dimension 2. We shall develop tools such as the graph of flats, which will lead us in Chap. 7 to results involving the geometric realization of the complex.

6.1 Basic Facts

In this chapter, we consider only simple paving simplicial complexes.

We start by establishing an alternative characterization of paving simplicial complexes using the lattice of flats:

Lemma 6.1.1. *Let* $\mathcal{H} = (V, H)$ *be a simplicial complex of dimension* d. *Then the following conditions are equivalent:*

(i) $\mathcal{H}$ *is paving;*
(ii) $P_{\leq d-1}(V) \subseteq \mathrm{Fl}\,\mathcal{H}$.

Proof. (i) $\Rightarrow$ (ii). By Proposition 4.2.2(iii).

(ii) $\Rightarrow$ (i). We show that $P_{\leq k}(V) \subseteq H$ for $k = 0, \ldots, d$ by induction. The case $k = 0$ being trivial, assume that $k \in \{1, \ldots, d\}$, $X \in P_{\leq k}(V)$ and $P_{\leq k-1}(V) \subseteq H$. Take $x \in X$. Then $X \setminus \{x\} \in P_{\leq k-1}(V) \subseteq H \cap \mathrm{Fl}\,\mathcal{H}$ and so $X \in H$. Thus $P_{\leq k}(V) \subseteq H$. By induction, we get $P_{\leq d}(V) \subseteq H$ and so $\mathcal{H}$ is paving. $\square$

Next we simplify the characterization of boolean representable paving simplicial complexes:

Proposition 6.1.2. *Let* $\mathcal{H} = (V, H) \in \mathrm{Pav}(d)$. *Then the following conditions are equivalent:*

© Springer International Publishing Switzerland 2015
J. Rhodes, P.V. Silva, *Boolean Representations of Simplicial Complexes and Matroids*, Springer Monographs in Mathematics,
DOI 10.1007/978-3-319-15114-4_6

(i) $\mathcal{H}$ *is boolean representable;*
(ii) $\forall X \in H \cap P_{d+1}(V) \; \exists x \in X : \quad x \notin \mathrm{Cl}(X \setminus \{x\})$;
(iii) $\forall X \in H \cap P_{d+1}(V) \; \exists x \in X : \quad \mathrm{Cl}(X \setminus \{x\}) \neq V$.

Proof. (i) $\Rightarrow$ (ii). By Theorem 5.2.6.

(ii) $\Rightarrow$ (iii). Immediate.

(iii) $\Rightarrow$ (i). By Corollary 5.2.7, it suffices to show that every $X \in H$ admits an enumeration $x_1, \ldots, x_k$ such that

$$\mathrm{Cl}(x_1, \ldots, x_k) \supset \mathrm{Cl}(x_2, \ldots, x_k) \supset \ldots \supset \mathrm{Cl}(x_k) \supset \mathrm{Cl}(\emptyset). \tag{6.1}$$

By condition (ii) in Lemma 6.1.1, this condition is satisfied if $|X| \leq d$. Thus we may assume that $|X| = d + 1$. By condition (iii), there exists some $x_1 \in X$ such that $\mathrm{Cl}(X \setminus \{x\}) \neq V$. By Proposition 4.2.3, we have $\mathrm{Cl}\, X = V$. Hence $\mathrm{Cl}\, X \supset \mathrm{Cl}(X \setminus \{x_1\})$. Now (6.1) follows easily from Lemma 6.1.1. $\square$

Since every matroid is pure, it is only natural to wonder which good properties pure paving simplicial complexes might possibly have. We close this section with a few counterexamples:

Example 6.1.3. Not every pure $\mathcal{H} = (V, H) \in \mathrm{Pav}(2)$ is boolean representable and the smallest counterexample occurs for $|V| = 5$.

Indeed, it follows easily from Example 5.2.11 that every pure $\mathcal{H} \in \mathrm{BPav}(2)$ is boolean representable when $|V| \leq 4$. Now let $V = \{1, \ldots, 5\}$ and $H = P_{\leq 2}(V) \cup \{123, 124, 125, 345\}$. It is immediate that $\mathcal{H}$ is paving and pure. Take $345 \in H$. It is easy to check that $1, 2 \in \overline{34}, \overline{35}, \overline{45}$ and so $\overline{34} = \overline{35} = \overline{45} = V$. In view of Proposition 6.1.2, $\mathcal{H}$ is not boolean representable.

Example 6.1.4. Not every pure $\mathcal{H} = (V, H) \in \mathrm{BPav}(2)$ is a matroid and the smallest counterexample occurs for $|V| = 5$.

It follows easily from Example 5.2.11 that every pure $\mathcal{H} \in \mathrm{BPav}(2)$ is a matroid when $|V| \leq 4$. Now let $V = \{1, \ldots, 5\}$ and $H = P_{\leq 3}(V) \setminus \{134, 135\}$. It is immediate that $\mathcal{H}$ is paving and pure. It is easy to check that $12, 23, 45 \in \mathrm{Fl}\, \mathcal{H}$. Since every facet must contain one of these flats, $\mathcal{H}$ is boolean representable by Proposition 6.1.2. Finally, (EP) fails for $I = 145$ and $J = 13$ and so $\mathcal{H}$ is not a matroid.

6.2 Graphic Boolean Simplicial Complexes

A simplicial complex $\mathcal{H} = (V, H)$ is said to be *graphic boolean* if it can be represented by a boolean matrix M such that:

- M contains all possible rows with one zero;
- Each row of M has at most two zeroes.

It follows from Lemma 6.1.1 that $\mathcal{H} = U_{2,|V|}$ or $\mathcal{H} \in \mathrm{BPav}(2)$.

We can then of course represent $\mathcal{H}$ by a graph where the edges correspond to the flats of the matrix having precisely two elements. This construction will be generalized in Sect. 6.3 under the notation ΓM.

Proposition 6.2.1. *Let* $\mathcal{H} = (V, H) \in \mathrm{Pav}(2)$. *Then the following conditions are equivalent:*

   (i) $\mathcal{H}$ is graphic boolean;
   (ii) If $X \in H \cap P_3(V)$, then $X \setminus \{x\} \in \mathrm{Fl}\,\mathcal{H}$ for some $x \in X$;
   (iii) There exist no $abc \in H$ and $x, y, z \in V$ such that $abx, ayc, zbc \notin H$.

Proof. (i) $\Rightarrow$ (ii). Let $\mathcal{H}$ be represented by a boolean matrix M satisfying the conditions in the definition of graphic boolean. Suppose that $X \in H$. Then, permuting rows and columns if necessary, we may assume that M has a 3×3 submatrix of the form

$$\begin{array}{|ccc}
1 & 0 & 0 \\
? & 1 & 0 \\
? & ? & 1 \\
\hline
a & b & c
\end{array}$$

It follows that there is a row in M having zeroes precisely at columns b and c, and this implies that zbc is c-independent for every $z \in V \setminus \{b, c\}$. Hence $zbc \in H$ for every $z \in V \setminus \{b, c\}$ and so $X \setminus \{a\} \in \mathrm{Fl}\,\mathcal{H}$.

   (ii) $\Rightarrow$ (iii). Suppose that exist some $abc \in H$ and $x, y, z \in V$ such that $abx, ayc, zbc \notin H$. Then $ab, ac, bc \notin \mathrm{Fl}\,\mathcal{H}$.

   (iii) $\Rightarrow$ (i). Let $M = \mathrm{Mat}\,\mathcal{H}$ and $F = \{X \in \mathrm{Fl}\,\mathcal{H} \mid |X| \leq 2\}$. We claim that $M[F, V]$ is a boolean representation of $\mathcal{H}$.

   If $X \subseteq V$ is c-independent with respect to $M[F, V]$, it is so with respect to M by Lemma 2.2.3(i) and so $X \in H$ by Lemma 5.2.3.

   Conversely, assume that $abc \in H$. By (iii), we may assume that $abx \in H$ for every $x \in V \setminus \{a, b\}$ and so $ab \in \mathrm{Fl}\,\mathcal{H}$. Hence $ab \in F$ and since $P_{\leq 1}(V) \subseteq F$ by Lemma 6.1.1, it follows that abc is c-independent with respect to $M[F, V]$. The equivalence is of course immediate for smaller subsets, hence $M[F, V]$ is a boolean representation of $\mathcal{H}$ and so $\mathcal{H}$ is graphic boolean.$\square$

We provide next a series of examples and counterexamples involving graphic boolean simplicial complexes:

Example 6.2.2. Not every $\mathcal{H} = (V, H) \in \mathrm{BPav}(2)$ is graphic boolean and the smallest counterexample occurs for $|V| = 5$.

   It is a simple exercise to show that $\mathcal{H} \in \mathrm{BPav}(2)$ is graphic boolean if $|V| \leq 4$ (see Example 5.2.11). Let $V = \{1, \ldots, 5\}$ and

$$H = P_{\leq 3}(V) \setminus \{123, 145, 245, 345\}.$$

Then Fl $\mathcal{H}$ $=$ $P_{\leq 1}(V) \cup \{123, V\}$. For every $X \in P_3(V) \cap H$, we have $|X \cap 123| = 2$, say $X \cap 123 = ij$. Writing $X = ijk$, we get

$$V = \mathrm{Cl}\, X \supset 123 = \mathrm{Cl}(ij) \supset i = \mathrm{Cl}(i) \supset \emptyset = \mathrm{Cl}(\emptyset).$$

The cases $X \in H \setminus P_3(V)$ being immediate, it follows from Corollary 5.2.7 that $\mathcal{H}$ is boolean representable. It is not graphic boolean since it fails Proposition 6.2.1(ii).

Example 6.2.3. Not every graphic boolean simplicial complex $\mathcal{H} = (V, H)$ is a matroid and the smallest counterexample occurs for $|V| = 4$.

The case $|V| \leq 3$ is trivial. For $|V| = 4$, we consider Example 5.2.11(iii), where it is shown that T_2 is not a matroid and Fl $T_2 = P_{\leq 1}(V) \cup \{12, V\}$. Therefore T_2 is graphic boolean by Proposition 6.2.1.

Example 6.2.4. Not every matroid $\mathcal{H} = (V, H) \in \mathrm{Pav}(2)$ is graphic boolean and the smallest counterexample occurs for $|V| = 6$.

The case $|V| \leq 4$ is easy to check in view of Example 5.2.11. Assume next that $\mathcal{H}$ is a paving matroid of dimension 2 with $V = \{1, \ldots, 5\}$.

Suppose that $123 \in$ Fl $\mathcal{H}$. Then $124, 125, 134, 135, 234, 235 \in H$. Since $124, 134, 234, 45 \in H$, it follows easily from (EP) that at least two of the 3-sets $145, 245, 345$ belong to H. Hence $P_{\leq 3}(V) \setminus H$ has at most two elements. Since the elements abx, ayc, zbc in the statement of Proposition 6.2.1 must be all distinct, it follows that $\mathcal{H}$ is graphic boolean.

Hence we may assume that Fl $\mathcal{H}$ contains no 3-set. Suppose that $1234 \in$ Fl $\mathcal{H}$. Then $ab5 \in H$ for all distinct $a, b \in \{1, \ldots, 4\}$ and so $a5 \in$ Fl $\mathcal{H}$ for every $a \in \{1, \ldots, 4\}$. By Proposition 4.2.3, H contains no other 3-set, hence the flats $a5$ suffice to build a representation of $\mathcal{H}$ and so $\mathcal{H}$ is graphic boolean.

Therefore we may assume that Fl $\mathcal{H}$ contains neither 3-sets nor 4-sets. Thus $\mathcal{H}$ is graphic boolean if $|V| = 5$.

Finally, let $V = \{1, \ldots, 6\}$ and $H = P_{\leq 3}(E) \setminus \{124, 135, 236\}$. Since $123 \in H$, it follows from Proposition 6.2.1 that $\mathcal{H}$ is not graphic boolean. However, $\mathcal{H} \in \mathrm{Pav}(2)$ and any two distinct elements of $P_{\leq 3}(V) \setminus H$ share precisely one element, so it follows easily that $\mathcal{H}$ is a matroid.

6.3   Computing the Flats in Dimension 2

We develop in this section techniques to compute the lattice of flats of a given $\mathcal{H} \in \mathrm{BPav}(2)$ from a given boolean representation.

We introduce the definition of *partial Euclidean geometry* (abbreviated to PEG). This concept has been extensively studied in a number of contexts in incidence geometry, incidence structures and set intersection problems. For example, see [10, 14, 38, 47, 48]. For the purposes of this book, it is convenient for us to call these structures PEGs.

Given a finite nonempty set V and a nonempty subset $\mathcal{L}$ of 2^V, we say that $(V, \mathcal{L})$ is a PEG if the following axioms are satisfied:

(G1) If $L, L' \in \mathcal{L}$ are distinct, then $|L \cap L'| \leq 1$;
(G2) $|L| \geq 2$ for every $L \in \mathcal{L}$.

The elements of V are called *points* and the elements of $\mathcal{L}$ are called *lines*.

Let $\mathcal{H} = (V, H) \in \mathrm{BPav}(2)$ be represented by an $R \times V$ boolean matrix $M = (m_{rp})$. By Lemma 5.2.1, $Z_r = \{p \in V \mid m_{rp} = 0\}$ belongs to Fl $\mathcal{H}$ for every $r \in R$. We say that Z_r is a *line* of M if $2 \leq |Z_r| < |V|$. We denote by $\mathcal{L}_M$ the set of lines of M and write

$$\mathrm{Geo}\, M = (V, \mathcal{L}_M).$$

Proposition 6.3.1. *Let M be a matrix representation of $\mathcal{H} \in \mathrm{BPav}(2)$. Then* $\mathrm{Geo}\, M$ *is a PEG.*

Proof. Since M has a submatrix congruent to

$$\begin{pmatrix} 1 & 0 & 0 \\ ? & 1 & 0 \\ ? & ? & 1 \end{pmatrix}$$

then $\mathcal{L}_M$ is nonempty. Now it suffices to prove axiom (G1). Suppose that $|L \cap L'| > 1$ for some distinct $L, L' \in \mathcal{L}_M$. We may assume that $L \nsubseteq L'$. Let $p \in L \cap L'$. By Proposition 4.2.2,

$$\emptyset \subset \{p\} \subset L \cap L' \subset L \subset V$$

is a chain in Fl $\mathcal{H}$. Since ht Fl $\mathcal{H} = 3$ by Corollary 5.2.8, we reach a contradiction. Therefore (G1) holds and $\mathrm{Geo}\, M$ is a PEG. $\square$

Assume now that $\mathcal{L} \subseteq 2^V$ is nonempty (not satisfying necessarily axioms (G1) or (G2)). We say that $X \subseteq V$ is a *potential line* with respect to $\mathcal{L}$ if $|X \cap L| \leq 1$ for every $L \in \mathcal{L} \setminus \{X\}$. We denote by $Po(\mathcal{L})$ the set of all potential lines with respect to $\mathcal{L}$, and by $Pom(\mathcal{L})$ the set of maximal elements of $Po(\mathcal{L})$ (with respect to inclusion).

Let $\mathcal{H} = (V, H)$ be a simplicial complex. The *rank function* $r_H : 2^V \to \mathbb{N}$ is defined by

$$X r_H = \max\{|I| \mid I \in 2^X \cap H\}.$$

The maximum value of r_H is the *rank* of $\mathcal{H}$ and equals $\dim \mathcal{H} + 1$. We collect more detailed information on rank functions in Sect. A.6 of the Appendix.

We prove now the following lemma:

Lemma 6.3.2. *Let M be a boolean representation of $\mathcal{H} = (V, H) \in \mathrm{BPav}(2)$. Then:*

(i) $\mathcal{L}_M \subseteq Pom(\mathcal{L}_M)$;

(ii) $Po(Pom(\mathcal{L}_M)) \subseteq Pom(\mathcal{L}_M) \cup P_{\leq 1}(V)$;

(iii) If $X \subset Y \in Po(\mathcal{L}_M) \setminus \mathcal{L}_M$, then $X \in Po(\mathcal{L}_M)$;

(iv) $X r_H \leq 2$ for every $X \in Po(\mathcal{L}_M)$.

Proof. (i) We have $\mathcal{L}_M \subseteq Po(\mathcal{L}_M)$ by Proposition 6.3.1. Suppose that $L \subset X$ with $L \in \mathcal{L}_M$ and $X \in Po(\mathcal{L}_M)$. Then $|X \cap L| = |L| > 1$. Since $X \neq L$, this contradicts $X \in Po(\mathcal{L}_M)$. Thus $\mathcal{L}_M \subseteq Pom(\mathcal{L}_M)$.

(ii) It follows from part (i) that $Po(Pom(\mathcal{L}_M)) \subseteq Po(\mathcal{L}_M)$. Suppose therefore that $X \in Po(Pom(\mathcal{L}_M))$, $|X| > 1$ and $X \subset Y$ for some $Y \in Po(\mathcal{L}_M)$. Then we may assume that $Y \in Pom(\mathcal{L}_M)$. Since $|X \cap Y| = |X| > 1$, this contradicts $X \in Po(Pom(\mathcal{L}_M))$. Thus $X \in Pom(\mathcal{L}_M)$.

(iii) Suppose that $|X \cap L| > 1$ for some $L \in \mathcal{L}_M \setminus \{X\}$. Then $|Y \cap L| > 1$ and $Y \notin \mathcal{L}_M$ yields $L \neq Y$. Thus $Y \notin Po(\mathcal{L}_M)$, a contradiction.

(iv) Suppose that $I \in H \cap 2^X$ with $|I| = 3$, say $I = abc$. After possible reordering of rows and columns, M has a submatrix of the form

$$
\begin{array}{c|ccc}
r_1 & 1 & 0 & 0 \\
r_2 & ? & 1 & 0 \\
r_3 & ? & ? & 1 \\
\hline
 & a & b & c
\end{array}
\tag{6.2}
$$

It follows that $b, c \in X \cap Z_{r_1}$, contradicting $X \in Po(\mathcal{L}_M)$. □

In the following lemma, we show how to recover H from $\mathcal{L}_M$ and $Pom(\mathcal{L}_M)$:

Lemma 6.3.3. *Let M be a boolean representation of $\mathcal{H} = (V, H) \in$ BPav(2). Then*

$$
H = P_{\leq 2}(V) \cup \left(\bigcup_{L \in \mathcal{L}_M} \{X \in P_3(V) \mid |X \cap L| = 2\} \right)
$$

$$
= P_{\leq 3}(V) \setminus \bigcup_{Y \in Pom(\mathcal{L}_M)} P_3(Y).
$$

Proof. Let $X \in H \setminus P_{\leq 2}(V)$. Then $|X| = 3$. Since X is c-independent with respect to M, there exists some submatrix of M of the form (6.2) with $X = abc$. Then $Z_{r_1} \in \mathcal{L}_M$ satisfies $|X \cap Z_{r_1}| = 2$ and so $H \subseteq P_{\leq 2}(V) \cup (\bigcup_{L \in \mathcal{L}_M} \{X \in P_3(V) \mid |X \cap L| = 2\})$.

Next, assume that $X \in P_3(V)$ satisfies $|X \cap L| = 2$ for some $L \in \mathcal{L}_M$. Suppose that $X \subseteq Y$ for some $Y \in Pom(\mathcal{L}_M)$. Then $|Y \cap L| \geq 2$, and $Y \in Po(\mathcal{L}_M)$ yields $Y = L$. Thus $|X| = |X \cap Y| = 2$, a contradiction. Therefore $P_{\leq 2}(V) \cup (\bigcup_{L \in \mathcal{L}_M} \{X \in P_3(V) \mid |X \cap L| = 2\}) \subseteq P_{\leq 3}(V) \setminus \bigcup_{Y \in Pom(\mathcal{L}_M)} P_3(Y)$.

Finally, assume that $X \in P_{\leq 3}(V) \setminus H$. Then $|X| = 3$, hence we must show that $X \subseteq Y$ for some $Y \in Pom(\mathcal{L}_M)$.

Suppose that $|X \cap L| > 2$ for some $L \in \mathcal{L}_M$. Then $X \subseteq L \in Pom(\mathcal{L}_M)$ by Lemma 6.3.2(i) and we are done. On the other hand, if $|X \cap L| = 2$ for some $L \in \mathcal{L}_M$, say $X \cap L = bc$, we use the fact that the b and c columns in M must be different (otherwise $bc \in H$ is not c-independent) to build a submatrix of M of the form (6.2), contradicting $X \notin H$. Hence we may assume that $|X \cap L| \leq 1$ for every $L \in \mathcal{L}_M$, i.e. $X \in Po(\mathcal{L}_M)$. Taking $Y \in Pom(\mathcal{L}_M)$ containing X, we reach our goal.$\square$

Now we use the operators P and Pom to describe Fl $\mathcal{H}$ from $\mathcal{L}_M$:

Theorem 6.3.4. *Let M be a boolean representation of $\mathcal{H} = (V, H) \in$ BPav(2). Then*

$$\text{Fl } \mathcal{H} = P_{\leq 1}(V) \cup \{V\} \cup Po(Pom(\mathcal{L}_M)).$$

Proof. Let $X \in$ Fl $\mathcal{H}$. We may assume that $1 < |X| < |V|$. Suppose that $X \notin Po(\mathcal{L}_M)$. Then $|X \cap L| > 1$ for some $L \in \mathcal{L}_M \setminus \{X\}$. Since $\mathcal{L}_M \subseteq$ Fl $\mathcal{H}$ by Lemma 5.2.1, this contradicts Proposition 6.3.1 (applied to the matrix Mat $\mathcal{H}$). Hence $X \in Po(\mathcal{L}_M)$.

Suppose that $|X \cap Y| > 1$ for some $Y \in Pom(\mathcal{L}_M) \setminus \{X\}$. Then $X \in Po(\mathcal{L}_M)$ yields $Y \not\subset X$. Take $a, b \in X \cap Y$ distinct and $p \in Y \setminus X$. Since $ab \in H$ and X is closed, we have $abp \in H$ and so $Yr_H > 2$, contradicting Lemma 6.3.2(iv). Therefore $X \in Po(Pom(\mathcal{L}_M))$.

Regarding the opposite inclusion, we have $P_{\leq 1}(V) \cup \{V\} \subseteq$ Fl $\mathcal{H}$ by Proposition 4.2.2. Let $X \in Po(Pom(\mathcal{L}_M))$ and assume that $I \in H \cap 2^X$ and $p \in V \setminus X$. We must show that $I \cup \{p\} \in H$.

Since $P_{\leq 2}(V) \subset H$, and in view of Lemma 6.3.2, we may assume that $|I| = 2$, say $I = ab$. Suppose that $abp \subseteq Y$ for some $Y \in Pom(\mathcal{L}_M)$. Since $p \notin X$, we have $X \neq Y$. However, $|X \cap Y| \geq 2$, contradicting $X \in Po(Pom(\mathcal{L}_M))$.

Thus no element of $Po(\mathcal{L}_M)$ contains abp. In particular, $abp \notin Po(\mathcal{L}_M)$ implies that there exists some $L \in \mathcal{L}_M \setminus \{abp\} \subseteq$ Fl $\mathcal{H}$ such that $|L \cap abp| > 1$. Since $\mathcal{L}_M \subseteq Po(\mathcal{L}_M)$ by Lemma 6.3.2(i), we cannot have $abp \subseteq L$, hence $|L \cap abp| = 2$. Hence, taking $x \in L \cap abp$, it follows that abp is a transversal of the successive differences for the chain $\emptyset \subset \{x\} \subset L \subset V$ in Fl $\mathcal{H}$, and so $abp \in H$ by Theorem 5.2.6. Therefore $X \in$ Fl $\mathcal{H}$.$\square$

Example 6.3.5. Let $V = \{1, \ldots, 7\}$ and $\mathcal{H} = (V, H)$ be represented by

$$M = \begin{pmatrix} 0 & 0 & 1 & 1 & 0 & 1 & 1 \\ 0 & 1 & 0 & 1 & 1 & 1 & 0 \\ 1 & 1 & 1 & 1 & 0 & 0 & 0 \end{pmatrix}$$

We compute Fl $\mathcal{H}$.

Note that M has no zero columns (hence every point is c-independent) and all columns are different (hence all 2-sets are c-independent). Since 123 is also c-independent, we have $\mathcal{H} \in \mathrm{BPav}(2)$. We generalize this example in Question 9.2.5.

If we represent Geo M as lines in the real plane,

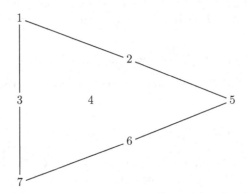

it is easy to see that $Pom(\mathcal{L}_M) = \mathcal{L}_M \cup \{146, 247, 345, 2346\}$. Since 2346 shares at least two points with each element of $\{146, 247, 345, 2346\}$, it follows that $Po(Pom(\mathcal{L}_M)) = \mathcal{L}_M$ and so Fl $\mathcal{H} = P_{\leq 1}(V) \cup \{V\} \cup \mathcal{L}_M$ by Theorem 6.3.4. Since dim $\mathcal{H} = 2$, any representation must have degree ≥ 3, hence mindeg $\mathcal{H} = 3$. Note also that by Lemma 6.3.3 H contains all 3-sets except those contained in some element of $Pom(\mathcal{L}_M)$. Thus

$$H = P_{\leq 3}(V) \setminus \{125, 137, 146, 234, 236, 246, 247, 345, 346, 567\}.$$

Given an $R \times V$ matrix M, we define a (finite undirected) graph $\Gamma M = (V, E)$, where E contains all edges of the form $p - q$ such that $p \neq q$ and $pq \subseteq Z_r$ for some $r \in R$. When M is a boolean representation of $\mathcal{H}$, ΓM can be of assistance on the computation of Fl $\mathcal{H}$ from $\mathcal{L}_M$.

Theorem 6.3.6. *Let M be a boolean representation of $\mathcal{H} = (V, H) \in \mathrm{BPav}(2)$. Then:*

$$\text{Fl } \mathcal{H} = P_{\leq 1}(V) \cup \{V\} \cup \mathcal{L}_M \cup \{\text{superanticliques of } \Gamma M\}.$$

Proof. By Lemma 5.2.1 and Theorem 6.3.4, we have $P_{\leq 1}(V) \cup \{V\} \cup \mathcal{L}_M \subseteq \text{Fl } \mathcal{H}$. Assume now that X is a superanticlique of ΓM. Let $I \in H \cap 2^X$ and $p \in V \setminus X$. We must show that $I \cup \{p\} \in H$.

Suppose that $|I| = 3$. By the first equality in Lemma 6.3.3, we have $|I \cap L| = 2$ for some $L \in \mathcal{L}_M$, contradicting X being an anticlique of ΓM. Hence $|I| \leq 2$, and we may indeed assume that $|I| = 2$, say $I = xy$. Since X is a superanticlique of ΓM, we have $p \in \mathrm{nbh}(x) \cup \mathrm{nbh}(y)$, hence we may assume that $p \in \mathrm{nbh}(x)$. It follows that $px \subseteq L$ for some $L \in \mathcal{L}_M$. Since $y \notin \mathrm{nbh}(x)$, we have $y \notin L$ and so $xyp \in H$ by the first equality in Lemma 6.3.3. Therefore $X \in \text{Fl } \mathcal{H}$.

Conversely, let $X \in \mathrm{Fl}\,\mathcal{H}$. We may assume that $1 < |X| < |V|$. Suppose first that there exists an edge $x \longrightarrow y$ in ΓM for some $x, y \in X$. Then $xy \subseteq L$ for some $L \in \mathcal{L}_M \subseteq \mathrm{Fl}\,\mathcal{H}$. Since $|X \cap L| > 1$, it follows from Proposition 6.3.1 (applied to the matrix Mat $\mathcal{H}$) that $X = L \in \mathcal{L}_M$. Hence we may assume that X is an anticlique. Let $x, y \in X$ be distinct and let $p \in V \setminus X$. Since $xy \in H$ and X is closed, we have $xyp \in H$. By the first equality in Lemma 6.3.3, xyp is not an anticlique. Since $x \notin \mathrm{nbh}(y)$, we get $p \in \mathrm{nbh}(x) \cup \mathrm{nbh}(y)$ and so $V \setminus X \subseteq \mathrm{nbh}(x) \cup \mathrm{nbh}(y)$. The opposite inclusion holds trivially in the anticlique X, hence X is a superanticlique. $\square$

We remark that, being easier for our eyes to detect cliques than anticliques, it is often useful in practice to work within $(\Gamma M)^c$ to exchange superanticliques by supercliques.

Next we combine Lemma 6.3.3 with Theorem 6.3.6 to obtain a new description of H in terms of M:

Proposition 6.3.7. *Let M be a boolean representation of $\mathcal{H} = (V, H) \in \mathrm{BPav}(2)$. Then*

$$H = P_{\leq 3}(V) \setminus (\{3\text{-anticliques of } \Gamma M\} \cup \bigcup_{L \in \mathcal{L}_M} P_3(L)).$$

Proof. Let $X \in P_3(V)$. By the second equality in Lemma 6.3.3, it suffices to show that $X \subseteq Y$ for some $Y \in Pom(\mathcal{L}_M)$ if and only if $X \in \cup_{L \in \mathcal{L}_M} P_3(L)$ or X is an anticlique of ΓM.

Assume first that $X \subseteq Y$ for some $Y \in Pom(\mathcal{L}_M)$ and X is not an anticlique of ΓM. The latter implies $|X \cap L| \geq 2$ for some $L \in \mathcal{L}_M$ and so also $|Y \cap L| \geq 2$. Hence $Y = L$ and so $X \in P_3(L)$.

For the opposite inclusion, in view of Lemma 6.3.2(i), we may assume that X is an anticlique of ΓM. Then $X \in Po(\mathcal{L}_M)$ and so is contained in some $Y \in Pom(\mathcal{L}_M)$. $\square$

Example 6.3.8. Let $V = \{1, \ldots, 5\}$ and $\mathcal{H} = (V, H)$ be represented by

$$M = \begin{pmatrix} 0 & 1 & 0 & 1 & 0 \\ 1 & 0 & 0 & 0 & 1 \\ 0 & 0 & 1 & 1 & 1 \end{pmatrix}$$

Then M has minimum degree among the representations of $\mathcal{H}$ and $\mathrm{Fl}\mathcal{H} = P_{\leq 1}(V) \cup \{V\} \cup \mathcal{L}_M \cup \{14, 25, 45\}$.

Note that all the columns are nonzero and different, and it follows easily that $\mathcal{H} \in \mathrm{BPav}(2)$. We cannot represent dimension 2 with two rows, hence M is a representation of minimum degree.

Now $(\Gamma M)^c$ is the graph

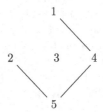

Clearly, the supercliques of $(\Gamma M)^c$ are $14, 25$ and 45. By Theorem 6.3.6, we get
$\mathrm{Fl}\,\mathcal{H} = P_{\leq 1}(V) \cup \{V\} \cup \mathcal{L}_M \cup \{14, 25, 45\}$.

By Proposition 6.3.7, we get also $H = P_{\leq 3}(V) \setminus \{135, 234\}$.

6.4 The Graph of Flats in Dimension 2

We explore in this section the concept of *graph of flats*. Recalling the definition of
ΓM in Sect. 6.3, we define $\Gamma\mathrm{Fl}\,\mathcal{H} = \Gamma\mathrm{Mat}\,\mathcal{H}$ for every $\mathcal{H} = (V, H) \in \mathrm{Pav}(2)$.
Thus we may write $\Gamma\mathrm{Fl}\,\mathcal{H} = (V, E)$, where p — q is an edge in E if and only if
$p \neq q$ and $\overline{pq} \subset V$.

In Chap. 7, the graph of flats will play a major role in the computation of the
homotopy type of a simplicial complex in $\mathrm{BPav}(2)$, namely through the number and
nature of its connected components.

We can characterize simple matroids through the graph of flats:

Proposition 6.4.1. *Let $\mathcal{H} \in \mathrm{Pav}(2)$. Then the following conditions are equivalent:*

(i) $\mathcal{H}$ is a matroid;
(ii) $\Gamma\mathrm{Fl}\,\mathcal{H}$ is complete.

Proof. (i) $\Rightarrow$ (ii). Write $\mathcal{H} = (V, H)$. Let $x, y \in V$ be distinct. Since every
matroid is pure, there exists some $z \in V \setminus xy$ such that $xyz \in H$. It follows from
Proposition 4.2.5(ii) that $z \notin \overline{xy}$, hence there exists an edge x — y in $\Gamma\mathrm{Fl}\,\mathcal{H}$.
Therefore $\Gamma\mathrm{Fl}\,\mathcal{H}$ is complete.

(ii) $\Rightarrow$ (i). Suppose that (EP) fails for $I, J \in H$. Since $P_2(V) \subseteq H$, then $|I| = 3$.
Since $J \cup \{i\} \notin H$ for every $i \in I \setminus J$, we must have $I \subseteq \overline{J}$. Hence $\overline{J}$ contains a
facet and so $\overline{J} = V$ by Proposition 4.2.3. Since $|J| = 2$, it follows that $\Gamma\mathrm{Fl}\,\mathcal{H}$ is
not complete.$\square$

Given a graph $\Gamma = (V, E)$, we define two simplicial complexes $\mathcal{H}^1(\Gamma) = (V, H^1$
$(\Gamma))$ and $\mathcal{H}^0(\Gamma) = (V, H^0(\Gamma))$ by

$$H^1(\Gamma) = \{X \in P_{\leq 3}(V) \mid X \text{ is not a 3-anticlique of } \Gamma\},$$
$$H^0(\Gamma) = \{X \in P_{\leq 3}(V) \mid X \text{ is neither a 3-clique nor a 3-anticlique of } \Gamma\}.$$

Clearly, $H^0(\Gamma) \subseteq H^1(\Gamma)$. We shall see that $H^1(\Gamma)$ and $H^0(\Gamma)$ are maximum and minimum in some precise sense. Note also that different graphs may yield the same complex, even if they have at least 3 vertices, even if they are connected:

Example 6.4.2. We may have $\mathcal{H}^1(\Gamma) = \mathcal{H}^0(\Gamma) = \mathcal{H}^1(\Gamma') = \mathcal{H}^0(\Gamma')$ for non-isomorphic graphs Γ and Γ'.

Indeed, let Γ and Γ' be depicted by

Then $\mathcal{H}^1(\Gamma) = \mathcal{H}^0(\Gamma) = \mathcal{H}^1(\Gamma') = \mathcal{H}^0(\Gamma') = P_{\leq 3}(V)$ since there are neither 3-cliques nor 3-anticliques in either graph.

Lemma 6.4.3. *Let $\mathcal{H} = (V, H) \in \mathrm{Pav}(2)$ and let $\Gamma = \Gamma\mathrm{Fl}\,\mathcal{H}$. Then $H^0(\Gamma) \subseteq H$.*

Proof. Write $\Gamma = (V, E)$. Let $X \in H^0(\Gamma)$. Since $\mathcal{H}$ is simple, we may assume that $X = abc$ with $ab \in E$ and $ac \notin E$. Hence $\overline{ab} \subset V = \overline{ac}$ and so $c \notin \overline{ab}$. Since $ab \in H$, we get $X \in H$.$\square$

The following lemma is essentially a restatement of Proposition 6.1.2:

Lemma 6.4.4. *Let $\mathcal{H} = (V, H) \in \mathrm{Pav}(2)$ and let $\Gamma = \Gamma\mathrm{Fl}\,\mathcal{H}$. Then the following conditions are equivalent:*

(i) $\mathcal{H} \in \mathrm{BPav}(2)$;
(ii) $H \subseteq H^1(\Gamma)$.

Proof. (i) $\Rightarrow$ (ii). Suppose that $H \nsubseteq H^1(\Gamma)$. Then there exists some $pqr \in H$ which is an anticlique. It follows that $\overline{pq} = \overline{pr} = \overline{qr} = V$, contradicting Proposition 6.1.2 since $\mathcal{H}$ is boolean representable. Therefore $H \subseteq H^1(\Gamma)$.

(ii) $\Rightarrow$ (i). Let $pqr \in H$. Then $pqr \in H^1(\Gamma)$ and so we may assume that $\overline{pq} \subset V$. If $r \in \overline{pq}$, then $\overline{pq}$ contains a facet of H and so $\overline{pq} = V$ by Proposition 4.2.3, a contradiction. Hence $r \notin \overline{pq}$ and so $\mathcal{H}$ is boolean representable by Proposition 6.1.2.$\square$

We now compute the flats for $\mathcal{H}^1(\Gamma)$ and $\mathcal{H}^0(\Gamma)$:

Proposition 6.4.5. *Let $\Gamma = (V, E)$ be a graph. Then:*

(i) $\mathrm{Fl}\,\mathcal{H}^1(\Gamma) = P_{\leq 1}(V) \cup \{V\} \cup E \cup \{\text{superanticliques of } \Gamma\}$;
(ii) $\mathrm{Fl}\,\mathcal{H}^0(\Gamma) = P_{\leq 1}(V) \cup \{V\} \cup \{\text{supercliques of } \Gamma\} \cup \{\text{superanticliques of } \Gamma\}$.

Proof. (i) We have $P_{\leq 1}(V) \cup \{V\} \subseteq \mathrm{Fl}\,\mathcal{H}^1(\Gamma)$ by Proposition 4.2.2. If $pq \in E$, then $pqr \in H^1(\Gamma)$ for every $r \in V \setminus pq$ and so $pq \in \mathrm{Fl}\,\mathcal{H}^1(\Gamma)$. Now assume that S is a superanticlique of Γ and let $I \in H^1(\Gamma) \cap 2^S$, $p \in V \setminus S$. The case $|I| = 3$ is impossible and $|I| \leq 1$ yields $I \cup \{p\} \in H^1(\Gamma)$ trivially. Hence we may assume that $|I| = 2$, say $I = ab$. But then $p \in \text{nbh}(a) \cup \text{nbh}(b)$ and so $I \cup \{p\} \in H^1(\Gamma)$. Therefore $S \in \text{Fl } \mathcal{H}^1(\Gamma)$.

Conversely, let $X \in \text{Fl } \mathcal{H}^1(\Gamma)$ and assume that $X \notin P_{\leq 1}(V) \cup \{V\}$. Assume first that X is not an anticlique. Let $p, q \in X$ be such that $pq \in E$. Since $pq \in \text{Fl } \mathcal{H}^1(\Gamma)$, we have a chain

$$\emptyset \subset p \subset pq \subseteq X \subset V.$$

If $pq \subset X$, it follows that $M = \text{Mat } \mathcal{H}^1(\Gamma)$ possesses a 4×4 lower unitriangular matrix and so there exists a c-independent 4-set with respect to M. Hence $\dim \mathcal{H}^1(\Gamma) > 3$ by Lemma 5.2.3, a contradiction. Thus $X = pq \in E$.

Finally, assume that X is an anticlique. Let $a, b \in X$ be distinct and let $p \in V \setminus X$. Then $abp \in H^1(\Gamma)$ yields $p \in \text{nbh}(a) \cup \text{nbh}(b)$. Hence $V \setminus X \subseteq \text{nbh}(a) \cup \text{nbh}(b)$ and the opposite inclusion follows from X being an anticlique. Thus X is a superanticlique.

(ii) We adapt the proof of part (i). If $X \in P_{\leq 1}(V) \cup \{V\}$ or X is a superanticlique, we get $X \in \text{Fl } \mathcal{H}^0(\Gamma)$ by the same arguments. Assume now that S is a superclique and let $I \in H^0(\Gamma) \cap 2^S$, $p \in V \setminus S$. The case $|I| = 3$ is impossible and $|I| \leq 1$ yields $I \cup \{p\} \in H^0(\Gamma)$ trivially. Hence we may assume that $|I| = 2$, say $I = ab$. But then $p \notin \overline{\text{nbh}}(a) \cap \overline{\text{nbh}}(b)$ and so $I \cup \{p\} \in H^0(\Gamma)$. Therefore $S \in \text{Fl } \mathcal{H}^0(\Gamma)$.

The opposite inclusion is a straightforward adaptation of the analogous proof in (i). $\square$

Corollary 6.4.6. *Let $\Gamma = (V, E)$ be a graph. Then:*

(i) $\mathcal{H}^1(\Gamma)$ is boolean representable;
(ii) $\mathcal{H}^0(\Gamma)$ need not be boolean representable.

Proof. (i) Let $pqr \in H^1(\Gamma)$. Then we may assume that $pq \in E$. Since pq is closed by Proposition 6.4.5, it follows from Proposition 6.1.2 that $\mathcal{H}^1(\Gamma)$ is boolean representable.

(ii) Let Γ be the graph described by

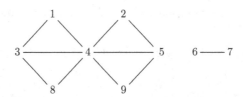

Then $345 \in H^0(\Gamma)$. Suppose that $\mathcal{H}^0(\Gamma)$ is boolean representable. By Proposition 6.1.2, we may assume without loss of generality that $5 \notin \overline{34}$ or $4 \notin \overline{35}$. The only maximal cliques containing 34 are 134 and 348, and it is easy to check that none of them is a superclique. Hence $\overline{34} = V$ by Proposition 6.4.5(ii). Similarly, the only maximal anticliques containing 35

are 356 and 357, and none of them is a superanticlique. Hence $\overline{35} = V$ by Proposition 6.4.5(ii). Thus we reach a contradiction and therefore $\mathcal{H}^0(\Gamma)$ is not boolean representable. $\square$

We shall give next abstract characterizations of $\Gamma\mathrm{Fl}\ \mathcal{H}$ for $\mathcal{H} \in \mathrm{Pav}(2)$ and $\mathcal{H} \in \mathrm{BPav}(2)$.

Theorem 6.4.7. *Let* $\Gamma = (V, E)$ *be a graph. Then* $\Gamma \cong \Gamma\mathrm{Fl}(V, H)$ *for some* $(V, H) \in \mathrm{Pav}(2)$ *if and only if the following conditions are satisfied:*

(i) $|V| \geq 3$;
(ii) *Every 2-anticlique of* Γ *is contained in some 3-anticlique;*
(iii) *For every 3-anticlique* X *of* Γ, *there exists some 3-anticlique* Y *such that* $|X \cap Y| = 2$.

Proof. Assume that $\Gamma = \Gamma\mathrm{Fl}(V, H)$ for some $(V, H) \in \mathrm{Pav}(2)$. Clearly, $|V| \geq 3$. Suppose that pq is an anticlique of Γ. Then pq is not closed and so $pqr \notin H$ for some $r \in V \setminus pq$. By Lemma 6.4.3, we get $pqr \notin H^0(\Gamma)$, hence pqr is an anticlique of Γ and (ii) holds.

Assume now that pqr is an anticlique of Γ. Suppose first that $pqr \in H$. Since pq is not closed, we have $pqs \notin H$ for some $s \in V \setminus pq$. Hence $s \neq r$. By Lemma 6.4.3, we get $pqs \notin H^0(\Gamma)$, hence pqs is an anticlique of Γ such that $|pqr \cap pqs| = 2$.

Thus we may assume that $pqr \notin H$, hence $|V| > 3$. Suppose pqr is closed. Since pqr is an anticlique, this implies $V = pqr$, a contradiction. Hence pqr is not closed, and so we have $xys \notin H$ for some $xy \subset pqr$ and $s \in V \setminus pqr$. By Lemma 6.4.3, xys is an anticlique of Γ such that $|pqr \cap xys| = 2$. Therefore (iii) holds.

Conversely, assume that conditions (i)–(iii) hold. Let S denote the set of all the 2-anticliques of Γ which belong to two different maximal anticliques. Assume that $X_1, \ldots, X_m$ are all the maximal anticliques of Γ having more than 3 elements and containing no element of S as a subset. For $i = 1, \ldots, m$, choose $Y_i \in P_3(X_i)$ and define

$$H = H^1(\Gamma) \cup \{Y_1, \ldots, Y_m\}.$$

Suppose that $E = \emptyset$. It follows from (iii) that $|V| > 3$. But then V is the unique maximal anticlique and has more than 3 elements, hence $Y_1 \in H$ and so $\dim(V, H) = 2$. The case $E \neq \emptyset$ is immediate, hence $(V, H) \in \mathrm{Pav}(2)$ in all cases.

Assume that $pq \in E$. Since $H^1(\Gamma) \subseteq H$, it follows from Proposition 6.4.5(i) that $pq \in \mathrm{Fl}(V, H)$ and so $p - q$ is an edge of $\Gamma\mathrm{Fl}(V, H)$.

Finally, assume that $pq \notin E$. We need to show that $\overline{pq} = V$. Suppose first that $pq \in S$. Then there exist two different maximal anticliques X and Y of Γ such that $pq \subseteq X \cap Y$. Let $x \in X \setminus pq$ and suppose that $pqx \in H$. Since X is an anticlique, we have $pqx \notin H^1(\Gamma)$. Hence $pqx = Y_i \subseteq X_i$ for some $i \in \{1, \ldots, m\}$, contradicting $pq \in S$. Hence $pqx \notin H$ and so $X \subseteq \overline{pq}$. Similarly, $Y \subseteq \overline{pq}$. Now, since $Y \neq X$ and X is a maximal anticlique, there exist some $x \in X$ and $y \in Y$ such that $xy \in E$. Hence $\overline{pq} \cap H^1(\Gamma)$ contains a 3-set and so $\overline{pq} = V$ by Proposition 4.2.3. This settles the case $pq \in S$.

Therefore we may assume that $pq \notin S$. Let X be the unique maximal anticlique of Γ containing pq. By (ii), we have $|X| \geq 3$. By (iii), and since $pq \notin S$, we have $|X| \geq 4$. Suppose that $pqx \notin H$ for every $x \in X \setminus pq$. Then $X \subseteq \overline{pq}$ and we may assume that $X = X_i$ for some $i \in \{1, \dots, m\}$, otherwise we use the case $pq \in S$ applied to some $p'q' \in P_2(X) \cap S$ to get $V = \overline{p'q'} \subseteq \overline{pq}$. But then $Y_i \subseteq \overline{pq}$ and so $\overline{pq} = V$ by Proposition 4.2.3.

Thus we may assume that there exists some $pqr \in H \cap 2^X$. Since X is an anticlique, we must have $pqr = Y_j \subseteq X_j$ for some $j \in \{1, \dots, m\}$. By uniqueness of X, we get $X = X_j$. Let $s \in X \setminus pqr$. We cannot have $pqs = Y_k$ for some k because $pq \notin S$. Since X is an anticlique, it follows that $pqs \notin H$ and so $s \in \overline{pq}$. If $qs \in S$, we may use the case $pq \in S$ applied to qs, hence we may assume that X is the unique maximal anticlique of Γ containing qs.

Suppose that $qrs \in H$. Since $qrs \notin H^1(\Gamma)$, we must have $qrs = Y_k \subseteq X_k$ for some $k \in \{1, \dots, m\}$. Since X_k is a maximal anticlique of Γ containing qs, we get $X_j = X = X_k$ and so $qrs = Y_k = Y_j = pqr$, a contradiction.

It follows that $qrs \notin H$ and so $r \in \overline{qs} \subseteq \overline{pq}$. Then $pqr \subseteq \overline{pq}$ and so $\overline{pq} = V$ by Proposition 4.2.3 as required.□

In particular, it follows that the square

is not of the form $\Gamma Fl(V, H)$ for some $(V, H) \in \mathrm{Pav}(2)$.

As we remarked before, going to the complement graph may make things easier, so we state the following corollary:

Corollary 6.4.8. *Let $\Gamma = (V, E)$ be a graph. Then $\Gamma \cong \Gamma Fl(V, H)$ for some $(V, H) \in \mathrm{Pav}(2)$ if and only if the following conditions are satisfied:*

(i) $|V| \geq 3$;
(ii) Every edge of Γ^c is contained in some triangle;
(iii) Every triangle of Γ^c shares exactly an edge with some other triangle.

We consider now the boolean representable case:

Theorem 6.4.9. *Let $\Gamma = (V, E)$ be a graph. Then $\Gamma \cong \Gamma Fl(V, H)$ for some $(V, H) \in \mathrm{BPav}(2)$ if and only if the following conditions are satisfied:*

(i) $|V| \geq 3$;
(ii) $E \neq \emptyset$;
(iii) Γ has no superanticliques.

Proof. Assume that $\Gamma = \Gamma\mathrm{Fl}(V, H)$ for some $(V, H) \in \mathrm{BPav}(2)$. Then there exists some 3-set $X \in H$ and so (i) holds. On the other hand, (ii) follows from Lemma 6.4.4.

Now assume that X is a nontrivial anticlique of Γ. Note that $X \subset V$ since $E \neq \emptyset$. However, since $|X| \geq 2$, we have $\overline{X} = V$ by definition of $\Gamma\mathrm{Fl}(V, H)$. It follows that X is not closed and so there exist some $I \in H \cap 2^X$ and $r \in V \setminus X$ such that $I \cup \{r\} \notin H$. Now, since (V, H) is simple, we must have $|I| \geq 2$. Since I is an anticlique, $|I| = 3$ contradicts Lemma 6.4.4, hence $|I| = 2$, say $I = pq$. Since $pqr \notin H$, we have $q \in \overline{pr}$ and so $\overline{pq} \subseteq \overline{pr}$. Since $q \notin \mathrm{nbh}(p)$, we have $\overline{pq} = V$, hence $\overline{pr} = V$. Similarly, $\overline{qr} = V$ and so pqr is a 3-anticlique of Γ. Thus $r \notin \mathrm{nbh}(p) \cup \mathrm{nbh}(q)$ and so X is not a superanticlique of Γ.

Conversely, assume that conditions (i)–(iii) do hold. By Corollary 6.4.6, $(V, H^1(\Gamma)) \in \mathrm{BPav}(2)$. By Proposition 6.4.5, and in view of (iii), we have

$$\mathrm{Fl}\,\mathcal{H}^1(\Gamma) = P_{\leq 1}(V) \cup \{V\} \cup E$$

and so $\Gamma\mathrm{Fl}(V, H^1(\Gamma)) = \Gamma$. $\square$

Corollary 6.4.10. *Let* $\Gamma = (V, E)$ *be a graph.*

(i) *If* $\Gamma \cong \Gamma\mathrm{Fl}(V, H)$ *for some* $(V, H) \in \mathrm{BPav}(2)$, *then* $H^1(\Gamma)$ *is the greatest possible* H *with this property with respect to inclusion.*

(ii) *If* Γ *is triangle-free and* $\Gamma \cong \Gamma\mathrm{Fl}(V, H)$ *for some* $(V, H) \in \mathrm{BPav}(2)$, *then* $H = H^1(\Gamma)$ *and* (V, H) *is graphic boolean.*

Proof. (i) By the proof of Theorem 6.4.9, together with Lemma 6.4.4.

(ii) By part (i), we have $H \subseteq H^1(\Gamma)$. Since Γ is triangle-free, $\overline{pq} = pq$ for every edge $p - q$ of Γ, and so $H^1(\Gamma) \subseteq H$. Moreover, every $F \in \mathrm{Fl}(V, H) \setminus \{V\}$ has at most two elements (to avoid triangles in Γ), thus (V, H) is graphic boolean. $\square$

The following example shows that the conditions of Theorems 6.4.7 and 6.4.9 are not equivalent, even if the graph has edges:

Example 6.4.11. We have $K_{1,4} \cong \Gamma\mathrm{Fl}(V, H)$ for some $(V, H) \in \mathrm{Pav}(2)$ but not for $(V, H) \in \mathrm{BPav}(2)$.

Next let $\mathcal{C}_m^n$ denote the class of all graphs having precisely n connected components and m nontrivial connected components.

In the disconnected case, we can get more precise characterizations than Theorem 6.4.9:

Proposition 6.4.12. *Let* $\Gamma = (V, E)$ *be a disconnected graph. Then* $\Gamma \cong \Gamma\mathrm{Fl}$ (V, H) *for some* $(V, H) \in \mathrm{BPav}(2)$ *if and only if* Γ *is not of the following types:*

(i) $\Gamma \in \mathcal{C}_0^n$ *for some* n;

(ii) $\Gamma \in \mathcal{C}_1^2$ with nontrivial connected component C, and C^c has a complete connected component;

(iii) $\Gamma \in \mathcal{C}_2^2$ with connected components C_1, C_2, and C_1^c, C_2^c both contain isolated points.

Proof. Assume that $\Gamma = \Gamma\mathrm{Fl}(V, H)$ for some $(V, H) \in \mathrm{BPav}(2)$. Then $|E| \neq \emptyset$ by Theorem 6.4.9 and so Γ is not type (i).

Suppose that Γ is type (ii). Let X be a complete connected component of C^c and write $V \setminus C = \{z\}$. Then $X \cup \{z\}$ is a nontrivial anticlique of Γ. Suppose that $a, b \in X \cup \{z\}$. Then $a \in C$ or $b \in C$ and so $\mathrm{nbh}(a) \cup \mathrm{nbh}(b) = V \setminus (X \cup \{z\})$. Hence $X \cup \{z\}$ is a superanticlique, contradicting Theorem 6.4.9. Thus Γ is not type (ii).

Suppose now that Γ is type (iii). Let z_i be an isolated point of C_i^c for $i = 1, 2$. Then $z_1 z_2$ is a superanticlique of Γ, contradicting Theorem 6.4.9, hence Γ is not type (iii) either.

Conversely, assume that Γ is neither type (i) nor type (ii) nor type (iii). We must show that the three conditions of Theorem 6.4.9 are satisfied. This is clear for the first two, so we suppose X to be a superanticlique of Γ.

Suppose first that Γ has at least three connected components. Since Γ is not type (i), has a nontrivial connected component C. Since X is a maximal anticlique, we can choose distinct $x, y \in X \setminus C$ and $z \in C \setminus X$. Then $z \notin \mathrm{nbh}(x) \cup \mathrm{nbh}(y)$, contradicting X being a superanticlique.

Suppose next that $\Gamma \in \mathcal{C}_2^2$ with connected components C_1, C_2. Since X is a maximal anticlique, it must intersect both C_1 and C_2. If $|X| = 2$, then the two elements of X must be isolated points of C_1^c and C_2^c, respectively. Since Γ is not type (iii), it follows that $|X| > 2$. Hence we may assume that there exist two distinct elements $x, y \in X \cap C_1$ and take $z \in C_2 \setminus X$. It follows that $z \notin \mathrm{nbh}(x) \cup \mathrm{nbh}(y)$, contradicting X being a superanticlique.

Therefore we may assume that $\Gamma \in \mathcal{C}_1^2$ with nontrivial connected component C and $V \setminus C = \{z\}$. Since Γ is not type (ii), $X \cap C$ cannot be a connected component of C^c (because it is a clique). Hence there exists an edge $x \text{ --- } y$ in C^c with $x \in X$ and $y \in C \setminus X$. It follows that $y \notin \mathrm{nbh}(x) \cup \mathrm{nbh}(z)$, contradicting X being a superanticlique.

Therefore Γ has no superanticliques and so $\Gamma \cong \Gamma\mathrm{Fl}(V, H)$ for some $(V, H) \in \mathrm{BPav}(2)$ by Theorem 6.4.9. $\square$

6.5 Computing mindeg $\mathcal{H}$ in Dimension 2

In this section, we compute mindeg $\mathcal{H}$ for every $\mathcal{H} \in \mathrm{BPav}(2)$ with $\Gamma\mathrm{Fl}\ \mathcal{H}$ disconnected.

Assuming $\mathcal{H} = (V, H)$ fixed, write $M = \mathrm{Mat}\ \mathcal{H}$. Let V_1 denote the set of points which belong to some single line $L \in \mathcal{L}_M$ (any line). Let also V_0 denote the set of points which belong to no line $L \in \mathcal{L}_M$. Note that V_0 consists of the isolated points in $\Gamma\mathrm{Fl}\ \mathcal{H}$, and

$$V_1 \subseteq \mathrm{smi}(\mathrm{Fl}\ \mathcal{H}).$$

We define

$$Q_0 = \begin{cases} 0 & \text{if } V_0 = \emptyset \\ 1 & \text{if } \Gamma\mathrm{Fl}\ \mathcal{H} \cong K_{m,1} \sqcup K_1 \\ |V_0| - 1 & \text{otherwise} \end{cases}$$

Write also

$$\mathcal{L}'_M = \{X \in \mathcal{L}_M \mid X \cap V_1 = \emptyset\}.$$

Theorem 6.5.1. *Let* $\mathcal{H} = (V, H) \in \mathrm{BPav}(2)$ *with* $\Gamma\mathrm{Fl}\ \mathcal{H}$ *disconnected. Let* $M = \mathrm{Mat}\ \mathcal{H}$. *Then*

$$\mathrm{mindeg}\ \mathcal{H} = |\mathcal{L}'_M| + |V_1| + Q_0.$$

Proof. Assume that N is a boolean matrix representation of $\mathcal{H}$ of minimum degree. By Theorem 5.2.5, we may assume that N is a submatrix of M. Thus $\mathcal{L}_N \subseteq \mathcal{L}_M$ and so ΓN is a subgraph of $\Gamma\mathrm{Fl}\ \mathcal{H}$ with the same vertex set V, hence disconnected. Suppose that $L \in \mathcal{L}_M \setminus \mathcal{L}_N$. Take $x, y \in L$ distinct. By Theorem 6.3.6 applied to N, we have $\mathrm{nbh}_{\Gamma M}(x) \cup \mathrm{nbh}_{\Gamma M}(y) = V \setminus L$ and so L intersects all the connected components of ΓN. Since ΓN is a subgraph of $\Gamma\mathrm{Fl}\ \mathcal{H}$, it follows that $\Gamma\mathrm{Fl}\ \mathcal{H}$ is connected, a contradiction. Hence $\mathcal{L}_M = \mathcal{L}_N$.

Since N contains no row of zeroes and $\mathrm{Fl}\ N \subseteq \mathrm{Fl}\ \mathcal{H}$, it follows that $\mathrm{Fl}\ N = \mathcal{L}_M \cup \mathcal{F}$ for some $\mathcal{F} \subseteq P_{\leq 1}(V)$. How small can $\mathcal{F}$ be? We start this discussion by decomposing V into a disjoint union

$$V = V_0 \cup V_1 \cup V_2.$$

Clearly, every $p \in V_2$ must belong to at least two lines L, L', and so $p = L \cap L'$ by Proposition 6.3.1. By minimality of N and Proposition 5.2.4(ii), we have $p \notin \mathrm{Fl}\ N$.

Let $L \in \mathcal{L}_M \setminus \mathcal{L}'_M$ and suppose that $p, q \in L \cap V_1$ are distinct. Since $pq \in H$ is c-independent with respect to N, and lines can't help to distinguish the two points, it follows that either p or q must belong to $\mathrm{Fl}\ N$. Hence $\mathrm{Fl}\ N$ must contain at least $|L \cap V_1| - 1$ flats of the form p $(p \in L \cap V_1)$. This implies the existence of $|V_1| - (|\mathcal{L}_M| - |\mathcal{L}'_M|)$ points of V_1 in $\mathcal{F}$.

Next suppose that $p, q \in V_0$ are distinct. Similar to the preceding case, either p or q must belong to $\mathrm{Fl}\ N$. Hence $\mathrm{Fl}\ N$ must contain at least $|V_0| - 1$ flats of the form p $(p \in V_0)$.

Finally, suppose that $\Gamma\mathrm{Fl}\ \mathcal{H} \cong K_{m,1} \sqcup K_1$. Let p be the vertex of degree m (note that $m > 1$ necessarily, otherwise $H = P_{\leq 3}(V)$ and $\Gamma\mathrm{Fl}\ \mathcal{H}$ is complete, a contradiction). Since $p \in H$ is c-independent with respect to N, and p belongs to every line, we must have $\emptyset$ or q in Fl N for some $q \neq p$. In any case, we may assume that $\mathcal{F}$ contains Q_0 flats of the form p ($p \in V_0$).

Write

$$K = |\mathcal{L}_M| + |V_1| - (|\mathcal{L}_M| - |\mathcal{L}'_M|) + Q_0 = |\mathcal{L}'_M| + |V_1| + Q_0.$$

All the above remarks combined show that

$$\text{mindeg } \mathcal{H} = |\text{Fl } N| \geq K$$

in all possible cases.

To prove the opposite inequality, we build a boolean representation N' of $\mathcal{H}$ with K rows. Indeed, let Fl N' contain:

(a) Every $L \in \mathcal{L}_M$;
(b) For every $L \in \mathcal{L}_M$ such that $|L \cap V_1| > 1$, all subsets of the form p ($p \in L \cap V_1$) but one;
(c) All subsets of the form p ($p \in V_0$) but one;
(d) $\emptyset$, if $\Gamma\text{Fl } \mathcal{H} \cong K_{m,1} \sqcup K_1$.

It is easy to check that Fl $N' \subseteq$ Fl $\mathcal{H}$ and $|\text{Fl } N'| = K$. It remains to show that every $X \in H$ is c-independent with respect to N'. We assume that the rows are indexed by the corresponding flats.

Assume first that $X = p$. We must show that $p \notin F$ for some $F \in$ Fl N'. In view of (a) and (c), we may assume that $p \in L$ for every $L \in \mathcal{L}_M$ and $|V_0| \leq 1$. Since $\Gamma\text{Fl } \mathcal{H}$ is disconnected, we must have indeed $|V_0| = 1$. Suppose that $|L'| > 2$ for some $L' \in \mathcal{L}_M$. Since $p \in L$ for every $L \in \mathcal{L}_M$ and by Proposition 6.3.1, it follows that $L \setminus \{p\} \subseteq V_1$ and so the claim follows from (b). Thus we may assume that $|L| = 2$ for every $L \in \mathcal{L}_M$. Therefore $\Gamma\text{Fl } \mathcal{H} \cong K_{m,1} \sqcup K_1$ and so $p \notin \emptyset \in$ Fl N'. This completes the case $|X| = 1$.

Assume next that $|X| = 2$, say $X = pq$. Suppose that there exists some $L \in \mathcal{L}_M$ such that $p \notin L$ and $q \in L$. Since there exists some $F \in$ Fl N' such that $q \notin F$, the submatrix $N[L, F; p, q]$ is of the form

$$
\begin{array}{c|cc}
L & 1 & 0 \\
F & ? & 1 \\
\hline
 & p & q
\end{array}
$$

and so pq is c-independent. Hence we may assume that p and q belong to the same lines. By Proposition 6.3.1, it follows that either $p, q \in V_0$ or $p, q \in L \cap V_1$ for some $L \in \mathcal{L}_M$. Using (c) or (b), respectively, and the case $|X| = 1$ as above, we complete the case $|X| = 2$.

Finally, the case $|X| = 3$ is proved using Lemma 6.3.3 (to get a row of the form 100) and the case $|X| = 2$ (to complete the construction of the lower unitriangular submatrix). Therefore every $X \in H$ is c-independent with respect to N' and so mindeg $\mathcal{H} = K$.□

Example 6.5.2. Let $V = \{1, \ldots, 6\}$. We compute mindeg $\mathcal{H}$ for the simplicial complex $\mathcal{H} = (V, H)$ represented by the matrix

$$M = \begin{pmatrix} 0\,0\,0\,1\,1\,1 \\ 1\,1\,1\,0\,0\,1 \\ 0\,1\,1\,1\,1\,1 \\ 1\,0\,1\,1\,1\,1 \\ 1\,1\,0\,1\,1\,1 \\ 1\,1\,1\,0\,1\,1 \\ 1\,1\,1\,1\,0\,1 \\ 1\,1\,1\,1\,1\,0 \end{pmatrix}$$

It is easy to check that $\mathcal{H} \in \mathrm{BPav}(2)$ and that the maximal potential lines are given by

$$Pom(\mathcal{L}_M) = \mathcal{L}_M \cup \{ab6 \mid a \in \{1, 2, 3\}, b \in \{4, 5\}\}.$$

Hence $Po(Pom(\mathcal{L}_M)) = \mathcal{L}_M$ and by Theorem 6.3.4 we get Fl $\mathcal{H} = $ Fl $M \cup \{V, \emptyset\}$. Hence ΓFl $\mathcal{H}$ is the graph

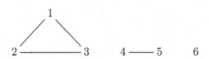

and is therefore disconnected. We may therefore apply Theorem 6.5.1:

We have $V_0 = \{6\}$, hence $Q_0 = 0$. Moreover, $V_1 = \{1, \ldots, 5\}$ since the two lines are disjoint, and so $\mathcal{L}'_M = \emptyset$. Thus Theorem 6.5.1 yields

$$\mathrm{mindeg}\ \mathcal{H} = |\mathcal{L}'_M| + |V_1| + Q_0 = 0 + 5 + 0 = 5.$$

It follows easily from the proof of Theorem 6.5.1 that

$$\begin{pmatrix} 0\,0\,0\,1\,1\,1 \\ 1\,1\,1\,0\,0\,1 \\ 0\,1\,1\,1\,1\,1 \\ 1\,0\,1\,1\,1\,1 \\ 1\,1\,1\,0\,1\,1 \end{pmatrix}$$

is a boolean representation of $\mathcal{H}$ with minimum degree.

Chapter 7
Shellability and Homotopy Type

In this section, we relate shellability of a simplicial complex $\mathcal{H} \in \mathrm{BPav}(2)$ with certain properties of its graph of flats. We then use shellability to determine the homotopy type of the geometric realization $\| \mathcal{H} \|$ (see Sect. A.5 in the Appendix) and compute its Betti numbers. We use the so-called nonpure version of shellability, introduced by Björner and Wachs in [5, 6].

7.1 Basic Notions

A simplicial complex $\mathcal{H} = (V, H)$ is *shellable* if we can order its facets as $B_1, \ldots, B_t$ so that, for $k = 2, \ldots, t$ and if $I(B_k) = (\cup_{i=1}^{k-1} 2^{B_i}) \cap 2^{B_k}$, then

$$(B_k, I(B_k)) \text{ is pure of dimension } |B_k| - 2 \tag{7.1}$$

whenever $|B_k| \geq 2$. Such an ordering is called a *shelling*. We say that B_k $(k > 1)$ is a *homology facet* in this shelling if $2^{B_k} \setminus \{B_k\} \subseteq \cup_{i=1}^{k-1} 2^{B_i}$.

Let $X_1, \ldots, X_n$ be mutually disjoint compact connected topological spaces. A *wedge* of $X_1, \ldots, X_n$, generically denoted by $\vee_{i=1}^{n} X_i$, is a topological space obtained by selecting a base point for each X_i and then identifying all the base points with each other. If each of the X_i has a transitive homeomorphism group (i.e., given any two points $x, y \in X_i$ there is a homeomorphism taling x into y), then $\vee_{i=1}^{n} X_i$ is unique up to homeomorphism. This is the case of spheres: a *sphere* of dimension d is a topological space homeomorphic to the Euclidean sphere

$$\{X \in \mathbb{R}^{d+1} \mid |X| = 1\},$$

where $|X|$ denotes the Euclidean norm of X.

© Springer International Publishing Switzerland 2015
J. Rhodes, P.V. Silva, *Boolean Representations of Simplicial Complexes and Matroids*, Springer Monographs in Mathematics,
DOI 10.1007/978-3-319-15114-4_7

Given topological spaces X and Y, a *homotopy* between continuous mappings $\phi, \psi : X \to Y$ is a family of continuous mappings $\theta_t : X \to Y$ ($t \in [0, 1]$) such that $\theta_0 = \varphi$, $\theta_1 = \psi$ and, for every $x \in X$, the mapping

$$[0, 1] \to Y$$
$$t \mapsto x\theta_t$$

is continuous for the usual topology of $[0, 1]$.

We say that two topological spaces have the same *homotopy type* if there exist continuous mappings $\alpha : X \to Y$ and $\beta : Y \to X$ such that:

- There exists a homotopy between $\alpha\beta$ and 1_X;
- There exists a homotopy between $\beta\alpha$ and 1_Y.

The homotopy type of a geometric simplicial complex (see Sect. A.5 in the Appendix) turns out to be undecidable in general, as we note in the end of this chapter, so the following theorem from Björner and Wachs illustrates the geometric importance of shellability. We omit defining some of the concepts appearing in it, and we omit the proof as well:

Theorem 7.1.1 ([5]). *Let $\mathcal{H}$ be a shellable trim simplicial complex of dimension d. Then:*

(i) $\| \mathcal{H} \|$ has the homotopy type of a wedge $W(\mathcal{H})$ of spheres of dimensions from 1 to d;

(ii) For $i = 1, \ldots, d$, the number $w_i(\mathcal{H})$ of i-spheres in the construction of $W(\mathcal{H})$ is the same as the following two numbers:

  – The number of homology facets of dimension i in a shelling of $\mathcal{H}$,
  – The i-th Betti number (i.e. the rank of the ith homology group) of $\| \mathcal{H} \|$.

For more details on shellability, the reader is referred to [53].

Our aim is to discuss shellability within Pav(2). We start with the low dimension cases which are easy to establish. The proof is left to the reader.

Proposition 7.1.2. *Let $\mathcal{H}$ be a simplicial complex of dimension ≤ 1. Then the following conditions are equivalent:*

(i) $\mathcal{H}$ is shellable;
(ii) There exists at most one nontrivial connected component of $(\Gamma \mathcal{H})^c$.

As preliminary work to the dimension 2 case, it is useful to establish when a simplicial complex with three vertices is pure of dimension 1. The proof is left to the reader.

Lemma 7.1.3. *Let $\mathcal{H} = (V, H)$ be a simplicial complex of dimension 1 with $|V| = 3$. Then $\mathcal{H}$ is pure if and only if one of the following conditions holds:*

(i) H contains exactly one 2-set but the third element of V is not in H;
(ii) H contains at least two 2-sets.

We may refer to simplicial complexes satisfying (i) (respectively (ii)) as *type 1* (respectively *type 2*).

The following result, due to Björner and Wachs [5, 6], shows that we may always rearrange the facets in a shelling with respect to dimension:

Lemma 7.1.4 ([5]). *Let $\mathcal{H}$ be a shellable simplicial complex. Then $\mathcal{H}$ admits a shelling where the dimension of the facets is not increasing.*

Proof. Let $B_1, \ldots, B_t$ be a shelling of $\mathcal{H}$. Suppose that $m = |B_i| < |B_{i+1}| = n$. We may assume that $m \geq 2$. We write $(B_j, I(B_j))$ with respect to the original shelling and $(B_j, I'(B_j))$ with respect to the sequence obtained by swapping B_i and B_{i+1}.

Let X be a facet of $(B_{i+1}, I'(B_{i+1}))$. Then $X \in I(B_{i+1})$. Since the complex $(B_{i+1}, I(B_{i+1}))$ is pure of dimension $n - 2$, we have $X \subseteq Y$ for some $Y \in I(B_{i+1})$ of dimension $n - 2$. Since $B_i \not\subseteq B_{i+1}$ and B_i has dimension $\leq n - 2$, we get $Y \in I'(B_{i+1})$ and so $(B_{i+1}, I'(B_{i+1}))$ is pure of dimension $n - 2$.

Now let X be a facet of $(B_i, I'(B_i))$. Suppose that $X \not\subseteq B_{i+1}$. Then $X \in I(B_i)$ and so $X \subseteq Y$ for some $Y \in I(B_i) \subseteq I'(B_i)$ of dimension $m - 2$. Hence we may assume that $X \subset B_{i+1}$. Since $(B_{i+1}, I(B_{i+1}))$ is pure of dimension $n - 2$, we have $X \subseteq Y$ for some $Y \in I(B_{i+1})$. Since $|Y| \geq m$ and $B_i \not\subseteq B_{i+1}$, it follows that $Y \in \cup_{j=1}^{i-1} 2^{B_j}$ and so $X \in I(B_i)$. Since $(B_i, I(B_i))$ is pure of dimension $m - 2$, we get $X \subseteq Z$ for some $Z \in I(B_i) \subseteq I'(B_i)$ of dimension $m - 2$. Therefore $(B_i, I'(B_i))$ is pure of dimension $m - 2$. Since $(B_j, I'(B_j)) = (B_j, I(B_j))$ for the remaining j, we still have a shelling after performing the swap. Performing all such swaps successively, we end up with a shelling of the desired type. $\square$

We end this section with a very useful result, due to Björner and Wachs [5], involving the notion of contraction. Given a simplicial complex $\mathcal{H} = (V, H)$ and $Q \in H \setminus \{V\}$, we define the *contraction* of $\mathcal{H}$ by Q to be the simplicial complex $(V \setminus Q, H/Q)$, where

$$H/Q = \{X \subseteq V \setminus Q \mid X \cup Q \in H\}.$$

When the simplicial complex $\mathcal{H}$ is implicit, this contraction is also known as the *link* of Q and denoted by $\mathrm{lk}(Q)$. See Sect. 8.3 for more details on contractions.

Proposition 7.1.5. *The class of shellable simplicial complexes is closed under contraction.*

Proof. Let $\mathcal{H} = (V, H)$ be a shellable simplicial complex with shelling $B_1, \ldots, B_t$ and let $A \in H \setminus \{V\}$. Given $X \subseteq V \setminus A$, we have $X \in H/A$ if and only if $X \cup A \in H$. Let $P = \{i \in \{1, \ldots, t\} \mid A \subseteq B_i\}$ and let $i_1, \ldots, i_m$ be the standard enumeration of the elements of P. It is straightforward to check that $B_{i_1} \setminus A, \ldots, B_{i_m} \setminus A$ constitutes an enumeration of the facets of $\mathrm{lk}(A)$. We prove that it is actually a shelling.

Let $k \in \{2, \ldots, m\}$. Then

$$I(B_{i_k} \setminus A) = (\cup_{j=1}^{k-1} 2^{B_{i_j} \setminus A}) \cap 2^{B_{i_k} \setminus A}$$

(with respect to the enumeration in $\mathrm{lk}(A)$). Suppose that X, Y are facets of $I(B_{i_k} \setminus A)$. Then $X \cup A \in I(B_{i_k})$ (with respect to the enumeration in $\mathcal{H}$) and it is easy to check that it is indeed a facet of $I(B_{i_k})$: if $X \cup A \subset X' \in I(B_{i_k})$, then $X \subset X' \setminus A \in I(B_{i_k} \setminus A)$, a contradiction. Similarly, also $Y \cup A$ is a facet of $I(B_{i_k})$. Since $(B_{I_k}, I(B_{i_k}))$ is pure, we get $|X \cup A| = |Y \cup A|$ and so also $|X| = |Y|$. Thus $(B_{i_k} \setminus A, I(B_{i_k} \setminus A))$ is pure. If $|B_{i_k} \setminus A| \geq 2$, it is easy to see that $(B_{i_k} \setminus A, I(B_{i_k} \setminus A))$ has dimension $|B_{i_k} \setminus A| - 2$, therefore $\mathrm{lk}(A)$ is shellable. $\square$

7.2   Shellability Within BPav(2)

We present in this section the main result of the whole chapter: the characterization of the shellable complexes within BPav(2) by means of the graph of flats $\Gamma\mathrm{Fl}\ \mathcal{H}$ defined in the beginning of Sect. 6.4. We prove a sequence of lemmas which, combined together, give the main theorem.

Let V be a finite set which we assume totally ordered. Let V^+ denote the set of all finite nonempty words (sequences) on V. Given two words $x = x_1 \ldots x_m$ and $y = y_1 \ldots y_n$ $(x_i, y_j \in V)$, we write $x < y$ if one of the following conditions is satisfied:

- There exists some $k \leq m, n$ such that $x_i = y_i$ for $1 \leq i < k$ and $x_k < y_k$;
- $m < n$ and $x_i = y_i$ for $1 \leq i \leq m$.

This is a total order known as the *alphabetic order* on V^*.

Given a simplicial complex $\mathcal{H} = (V, H)$ and a total order on V, we define the alphabetic order on fct $\mathcal{H}$ as follows. Given $B \in$ fct $\mathcal{H}$, let $\mathrm{ord}(B)$ denote the word of V^+ obtained by enumerating the elements of B in increasing order. Given $B, B' \in$ fct $\mathcal{H}$, we write

$$B < B' \quad \text{if } \mathrm{ord}(B) < \mathrm{ord}(B')$$

for the alphabetic order on V^+.

Lemma 7.2.1. *Let $\mathcal{H} \in$ BPav(2) with at most one nontrivial connected component in $\Gamma\mathrm{Fl}\ \mathcal{H}$. Then $\mathcal{H}$ is shellable.*

Proof. Write $\mathcal{H} = (V, H)$. Note that, in view of Theorem 6.4.9, $\Gamma = \Gamma\mathrm{Fl}\ \mathcal{H}$ has precisely one nontrivial connected component C. Consider Geo Mat $\mathcal{H} = \{F \in \mathrm{Fl}\ \mathcal{H} \mid 2 \leq |F| < |V|\}$ and write

$$\mathrm{Geo\,Mat}\ \mathcal{H} = \{F_1, \ldots, F_m\}.$$

Each F_i defines a clique in Γ, and these cliques cover C completely. Since C is connected, we may assume that $F_1, \ldots, F_m$ is an enumeration such that

$$F_i \cap (F_1 \cup \ldots \cup F_{i-1}) \neq \emptyset$$

for $i = 2, \ldots, m$. By Proposition 6.3.1, there exists a unique element

$$v_i \in F_i \cap (F_1 \cup \ldots \cup F_{i-1})$$

for $i = 2, \ldots, m$. We write also

$$F_i' = F_i \setminus (F_1 \cup \ldots \cup F_{i-1})$$

for $i = 1, \ldots, m$.

Now, if $F_i' \neq \emptyset$, we fix a total order on F_i' having v_i as minimum (if $i > 1$). We fix also an arbitrary total order on $V \setminus C$. We glue these total orders together according to the scheme

$$F_1' < F_2' < \ldots < F_m' < V \setminus C$$

to get a total order on V.

We consider now an enumeration $B_1, B_2, \ldots, B_n$ of the facets of $\mathcal{H}$ with respect to the alphabetic ordering. We claim that this enumeration is a shelling of $\mathcal{H}$.

Let $k \in \{2, \ldots, n\}$. Assume first that $|B_k| = 2$, say $B_k = xy$. Suppose that $x \in C$. If y is not adjacent to x, take $z \in V$ adjacent to x. Then $y \notin \overline{xz}$ and Lemma 6.4.3 yields $xyz \in H$, contradicting $B_k \in \text{fct } \mathcal{H}$. Hence we may assume that $x \relbar y$ is an edge of Γ. Taking $z \in V \setminus \overline{xy}$, once again Lemma 6.4.3 yields $xyz \in H$, contradicting $B_k \in \text{fct } \mathcal{H}$.

Therefore we must have $B_k \cap C = \emptyset$. Let $a \relbar b$ be an edge in C. Then $abx, aby \in \text{fct } \mathcal{H}$ in view of Lemma 6.4.3. Since $abx, aby < B_k$, it follows that $(B_k, I(B_k))$ is pure of dimension 0.

Hence we may assume that $|B_k| = 3$. Write $\text{ord}(B_k) = xyz$. By Lemma 6.4.4, we have $|B_k \cap C| \geq 2$. Hence $x \in F_i'$ for some $i \in \{1, \ldots, m\}$.

Assume first that $y \in F_i'$. Then $z \notin F_i$ in view of Proposition 4.2.3. Suppose that $i > 1$. Since $v_i \in F_i$, it follows from Lemma 6.4.3 that $v_0xz, v_0yz \in H$. Since $v_0 < x < y$, we get $v_0xz, v_0yz < B_k$ and so $(B_k, I(B_k))$ is pure of dimension 1 by Lemma 7.1.3.

Hence we may assume that $i = 1$. Let $p < q$ denote the first two elements of V. If $x \neq 1$, by adapting the preceding argument we get $pxz, pyz \in H$, $pxz, pyz < B_k$ and so $(B_k, I(B_k))$ is pure of dimension 1.

Thus we may assume that $x = p$. If $y \neq q$, we repeat the same argument using pqy, pqz. Therefore we may assume that $y = q$. But then the only elements which can precede B_k in the ordering of fct $\mathcal{H}$ are of the form pqr with $q < r < z$ and so $(B_k, I(B_k))$ is pure of dimension 1 also in this case.

Thus we may assume that $y \notin F_i'$. Then $y \in F_j'$ for some $j > i$. Then $z \in F_j$ by Lemma 6.4.4. Let $x' \in F_i \setminus \{x\}$. It is easy to check that $xx'y, xx'z \in H$, $xx'y, xx'z < B_k$ and so $(B_k, I(B_k))$ is pure of dimension 1 in this final case.

Therefore $B_1, B_2, \ldots, B_n$ is a shelling of $\mathcal{H}$. $\square$

The following example shows that the ordering of V cannot be arbitrary, even if $\Gamma\mathrm{Fl}\,\mathcal{H}$ is connected. This is in contrast with the case of matroids, where the shelling can be defined through any ordering of the vertices [4] (see also [53]).

Example 7.2.2. Let Γ be the graph described by

$$1 - 5 - 3 - 4 - 6 - 7 - 2$$

where the vertices are ordered by the usual integer ordering. Then the alphabetic order on the facets of $\mathcal{H}^1(\Gamma)$ does not produce a shelling.

First, we check that Γ has no superanticliques and so $\Gamma \cong \Gamma\mathrm{Fl}\,\mathcal{H}^1(\Gamma)$ by Theorem 6.4.9 and Corollary 6.4.10. Then we note that the alphabetic order on the facets starts with

$$125 < 127 < 134 < \ldots$$

and so $(134, I(134))$ is not pure of dimension 1.

The following example shows that Lemma 7.2.1 may fail if $\mathcal{H}$ is not boolean representable:

Example 7.2.3. Let $\mathcal{H} = (V, H)$ be defined by $V = \{1, \ldots, 5\}$ and $H = P_{\leq 2}(V) \cup \{123, 124, 125, 345\}$. Then:

  (i) $\mathcal{H} \in \mathrm{Pav}(2)$;
 (ii) $\mathcal{H}$ is not boolean representable;
(iii) $\Gamma\mathrm{Fl}\,\mathcal{H} \in \mathcal{C}_1^4$;
 (iv) $\mathcal{H}$ is not shellable.

It is straightforward to check that

$$\mathrm{Fl}\,\mathcal{H} = P_{\leq 1}(V) \cup \{12, V\}.$$

Since $345 \notin H^1(\Gamma)$, it follows from Lemma 6.4.4 that $\mathcal{H}$ is not boolean representable. Moreover, $\Gamma\mathrm{Fl}\,\mathcal{H}$ is the graph

$$1 - 2 \qquad 3 \qquad 4 \qquad 5$$

and is therefore in $\mathcal{C}_1^4$.

Clearly, $\mathrm{lk}(5) = (\{1, \ldots, 4\}, H')$ for $H' = P_{\leq 1}(1234) \cup \{12, 34\}$, hence $\mathrm{lk}(5)$ is not shellable by Proposition 7.1.2. By Proposition 7.1.5, $\mathcal{H}$ is not shellable either.

Lemma 7.2.4. *Let* $\mathcal{H} \in \mathrm{Pav}(2)$ *with* $\Gamma\mathrm{Fl}\,\mathcal{H} \in \mathcal{C}_m^m$ *and* $m \geq 1$. *Then* $\mathcal{H}$ *is pure.*

Proof. Write $\mathcal{H} = (V, H)$ and $\Gamma = \Gamma\mathrm{Fl}\,\mathcal{H} = (V, E)$. Suppose that pq is a facet of $\mathcal{H}$. If $pq \notin E$, we can take $x \in \mathrm{nbh}(p)$ since Γ has no trivial connected components. Then $pqx \in H^0(\Gamma) \subseteq H$ by Lemma 6.4.3. On the other hand, if $pq \in E$, we can take y in some other connected component and we also get $pqy \in H^0(\Gamma) \subseteq H$. Therefore $\mathcal{H}$ is pure. $\square$

Lemma 7.2.5. *Let* $\mathcal{H} \in \mathrm{Pav}(2)$ *with* $\Gamma\mathrm{Fl}\ \mathcal{H} \in \mathcal{C}_2^2$. *Then* $\mathcal{H}$ *is shellable.*

Proof. Write $\mathcal{H} = (V, H)$ and $\Gamma = \Gamma\mathrm{Fl}\ \mathcal{H} = (V, E)$. We consider first the particular case of $\mathcal{H}$ being boolean representable.

Let $V = A \cup B$ be the partition defined by the connected components. Write $A = \{a_1, \ldots, a_m\}$ and assume that, for every $i \in \{2, \ldots, m\}$, we have $a_i a_{i\alpha} \in E$ for some $i\alpha < i$. Such an enumeration exists because A is the set of vertices of a connected component. Similarly, we may write $B = \{b_1, \ldots, b_n\}$ and assume that, for every $j \in \{2, \ldots, n\}$, we have $b_j b_{j\beta} \in E$ for some $j\beta < j$. By Lemma 6.4.3, $a_i a_j b_k \in H$ whenever $a_i a_j \in E$. Similarly, $a_i b_j b_k \in H$ whenever $b_j b_k \in E$.

Suppose now that $a_i a_j a_k \in H$. By Lemma 6.4.4, at least two of these three vertices must be connected by some edge. Since they belong to the same connected component, and permuting i, j, k if necessary, there exists some path in Γ of the form

$$a_i \relbar a_j \relbar y_1 \relbar \ldots \relbar y_\ell \relbar a_k. \tag{7.2}$$

We denote by H_ℓ' the set of all $a_i a_j a_k$ such that there is a path of the form (7.2) in Γ and ℓ is minimal. Similarly, we define H_ℓ'' considering the facets $b_i b_j b_k$.

Consider the sequence of facets of (E, H)

$$a_2 a_{2\alpha} b_1, \ldots, a_m a_{m\alpha} b_1, \quad a_2 a_{2\alpha} b_2, \ldots, a_m a_{m\alpha} b_2,$$
$$\ldots, \quad a_2 a_{2\alpha} b_n, \ldots, a_m a_{m\alpha} b_n,$$

followed by successive enumerations of:

- The remaining facets of the form $a_i a_j b_k$,
- The facets of the form $a_i b_j b_k$,
- The facets of $H_0', H_1', \ldots,$
- The facets of $H_0'', H_1'', \ldots$.

In view of Lemma 7.2.4, it is easy to see that this is an enumeration of all the facets of $\mathcal{H}$. We claim it is indeed a shelling. In most of the instances, this involves straightforward checking that can be essentially omitted, hence we just focus on the hardest cases: the H_ℓ' (the H_ℓ'' cases are similar).

Suppose that $a_i a_j a_k \in H_0'$. We may assume that we have a path

$$a_i \relbar a_j \relbar a_k,$$

hence $a_i a_j b_1$ and $a_j a_k b_1$ appeared before and we may use Lemma 7.1.3.

Suppose now that $a_i a_j a_k \in H_\ell'$ with $\ell > 0$, and assume that (7.1) holds for all facets in $H_{\ell-1}'$. We may assume that there is a path in Γ of the form (7.2) with ℓ minimal.

Suppose that $a_j y_\ell a_k \notin H$. Since $\overline{y_\ell a_k}$ is closed, it follows that $a_j \in \overline{y_\ell a_k}$, hence $\overline{a_j a_k} \subseteq \overline{y_\ell a_k} \subset V$ (since $y_\ell a_k \in E$) and so there is an edge $a_j \relbar a_k$, contradicting $\ell > 0$. Thus $a_j y_\ell a_k \in H$.

Suppose now that $a_i a_j y_\ell \notin H$. Since $\overline{a_i a_j}$ is closed, it follows that $y_\ell \in \overline{a_i a_j}$, hence $\overline{a_j y_\ell} \subseteq \overline{a_i a_j} \subset V$ and so there is an edge $a_j \text{ --- } y_\ell$. By minimality of ℓ, it follows that $\ell = 1$. Moreover, $a_k \notin \overline{a_i a_j}$ (otherwise $\overline{a_i a_j} \subset V$ would contain a facet, contradicting Proposition 4.2.3), hence we get $a_k \notin \overline{a_j y_\ell}$ and so $a_j y_\ell a_k \in H$. Similarly, $\overline{a_i y_\ell} \subseteq \overline{a_i a_j}$ yields $a_k \notin \overline{a_i y_\ell}$ and so $a_i y_\ell a_k \in H$. Now note that $a_j y_\ell a_k \in H_0'$, and in fact also $a_i y_\ell a_k \in H_0'$ since $\overline{a_i y_\ell} \subseteq \overline{a_i a_j} \subset V$. Hence $a_i a_k, a_j a_k \in I(a_i a_j a_k)$ and so (7.1) holds for $a_i a_j a_k$ by Lemma 7.1.3 in this case.

Thus we may assume that $a_i a_j y_\ell \in H$. It follows that $a_j y_\ell a_k, a_i a_j y_\ell \in H_{\ell-1}'$, hence $a_i a_j, a_j a_k \in I(a_i a_j a_k)$ and so (7.1) holds for $a_i a_j a_k$ too in this case.

This completes the discussion of the crucial H_ℓ' cases and the proof of the boolean representable case.

We consider now the general case. Let $J = H \cap H^1(\Gamma)$. Then $(V, J) \in \text{Pav}(2)$. Write $\Gamma \text{Fl}(V, J) = \Gamma' = (V, E')$. We claim that Fl $\mathcal{H} \subseteq \text{Fl}(V, J)$. Indeed, let $X \in \text{Fl } \mathcal{H}$. We may assume that $X \subset V$. Let $I \in J \cap 2^X$ and $p \in V \setminus X$. Since $P_{\le 2}(V) \subseteq J$, we may assume that $|I| \ge 2$. On the other hand, $|I| = 3$ implies that X contains a facet of $\mathcal{H}$ and so $X = V$ by Proposition 4.2.3, a contradiction. Thus $|I| = 2$. Since $J \subseteq H$ and $X \in \text{Fl } \mathcal{H}$, we get $I \cup \{p\} \in H$. Now $I \subseteq X \in \text{Fl}$ $\mathcal{H} \setminus \{V\}$ implies that I is a clique in Γ, hence $I \cup \{p\} \in H^1(\Gamma)$ and so $I \cup \{p\} \in J$. Therefore Fl $\mathcal{H} \subseteq \text{Fl}(V, J)$.

It follows that $E \subseteq E'$ and so $J \subseteq H^1(\Gamma) \subseteq H^1(\Gamma')$. By Lemma 6.4.4, $(V, J) \in$ BPav(2). Since we have already proved the boolean representable case, we may assume $B_1, \ldots, B_t$ to be a shelling of (V, J). Clearly, every facet of dimension 2 of (V, J) is still a facet of $\mathcal{H}$. We enumerate the facets of $\mathcal{H}$ starting with $B_1, \ldots, B_t$ followed by the remaining facets $B_1', \ldots, B_m'$ in an arbitrary way. We claim this is a shelling of $\mathcal{H}$.

Indeed, by Lemma 7.2.4 we may write $B_i' = pqr$. Since $B_i' \notin J$, it is an anticlique of Γ. Let $x \in \text{nbh}(p)$ and $y \, \text{nbh}(q)$. Then $pxr, yqr \in H^0(\Gamma)$ and so $pxr, yqr \in H$ by Lemma 6.4.3. Since $H^0(\Gamma) \subseteq H^1(\Gamma)$, we get $pxr, yqr \in J$ and so $pr, qr \in I(B_i')$. Therefore (7.1) holds for B_i' and we have indeed a shelling. $\square$

The next example shows that, in general, we cannot assume that the shelling is defined by an alphabetic ordering, evidence of further deviation from the matroid case [4].

Example 7.2.6. Let Γ be the graph described by

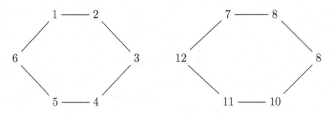

Then no shelling of $\mathcal{H}^1(\Gamma)$ can be defined through an alphabetic ordering of the facets.

First, we check that Γ has no superanticliques and so $\Gamma \cong \Gamma$Fl $\mathcal{H}^1 (\Gamma)$ by Theorem 6.4.9 and Corollary 6.4.10. Fix an ordering of the vertices. We may assume without loss of generality that 1 is the minimum element. Let $X = \{7, 8, \ldots, 12\}$. All facets of $\mathcal{H}$ have three elements. Note that if $B \in$ fct $\mathcal{H}$ contains precisely two elements of X, then these two vertices must be adjacent in Γ by definition of $\mathcal{H}^1 (\Gamma)$.

We may assume that $\min X = 7$ and $\min\{8, \ldots, 12\} \in \{8, 9, 10\}$ for our ordering. We split the discussion into these three cases.

Suppose first that $\min\{8, \ldots, 12\} = 8$. If $\{9, 10, 11, 12\}$ are ordered as $a < b < c < d$, then the first facets contained in X to appear in the alphabetic order are

$$\ldots < 78a < \ldots < 78b < \ldots < 78c < \ldots < 78d < \ldots$$

Let $B = \{7, 8, 10\}$. Now $I(B)$ contains 78 (since 178 is a facet and $178 < 78a$) but does not contain $\{7, 10\}$ because these two vertices are not adjacent and so could not have appeared as part of a facet containing an element of $\{1, \ldots, 6\}$. Similarly, $\{8, 10\} \notin I(B)$. However, $10 \in I(B)$ since the facet $\{1, 10, 11\}$ must have appeared before. Thus $(B, I(B))$ is not pure.

Suppose next that $\min\{8, \ldots, 12\} = 9$. Then the first facet contained in X to appear in the alphabetic order is of the form $79x$ for some $x \in \{8, 10, 12\}$ (since two of the vertices must be adjacent). It is easy to check that $(B, I(B))$ is not pure if $x \neq 8$, hence we may assume that the first facets contained in X to appear in the alphabetic order are:

- 798,
- $\{7, 9, 10\}$ and $\{7, 9, 12\}$ (in any order),
- $\{7, 8, 10\}$, $\{7, 8, 11\}$ and $\{7, 8, 12\}$ (in any order).

Let $B = \{7, 8, 12\}$. As in the preceding case, it is easy to check that $(B, I(B))$ is not pure.

Finally, suppose that $\min\{8, \ldots, 12\} = 10$. Then the first facet contained in X to appear in the alphabetic order is of the form $B = \{7, 10, x\}$ and it follows easily that $(B, I(B))$ is not pure. Therefore the alphabetic order never produces a shelling in this example.

Lemma 7.2.7. *Let $\mathcal{H} \in$ BPav(2) with ΓFl $\mathcal{H} \in \mathcal{C}_u^v$, $u \geq 2, v \geq 3$. Then $\mathcal{H}$ is not shellable.*

Proof. Write ΓFl $\mathcal{H} = \Gamma = (V, E)$. Let $V = V_1 \cup \ldots \cup V_m$ be the decomposition of V in its connected components.

Assume first that Γ has at least one trivial connected component. Assume that $V_1, \ldots, V_t$ are the nontrivial ones and write $V_m = \{v\}$. In view of Proposition 7.1.5, it suffices to show that $\mathrm{lk}(v)$ is not shellable.

Let $p, q \in V \setminus \{v\}$ be distinct. By Lemma 6.4.4, $pqv \in H$ implies $pq \in E$. Conversely, if $pq \in E$, then $v \notin \overline{pq} \subset V$ and so $pqv \in H$. Thus pq is a facet of lk(v) if and only if $pq \in E$. Since Γ has at least two trivial connected components, it follows that $(\Gamma \mathrm{lk}(v))^c$ has more than one nontrivial connected component. Thus lk(v) is not shellable by Proposition 7.1.2 and so is $\mathcal{H}$.

Assume now that Γ has no trivial connected components. We consider first the particular case in which each connected component of Γ is complete and $H = H^1(\Gamma)$. Note that, by Corollary 6.4.10 and Proposition 6.4.12, we have indeed $(V, H^1(\Gamma)) \in \mathrm{BPav}(2)$ for such a graph Γ. Moreover, all the facets have dimension 2 by Lemma 7.2.4.

Suppose that $B_1, \ldots, B_t$ is a shelling of $\mathcal{H}$. Let $B_{i_1}, B_{i_2}, \ldots, B_{i_{n-2}}$ denote the type 1 facets (recall the terminology introduced after Lemma 7.1.3). Clearly, three symbols have their first appearance in $B_{i_1} = B_1$, and then they appear one at the time in $B_{i_2}, \ldots, B_{i_{n-2}}$ (cf. Lemma 7.1.3). We build an enumeration of the facets through blocks $\mathcal{Q}_1, \ldots, \mathcal{Q}_{n-2}$ satisfying the following properties:

- $\mathcal{Q}_1$ contains only the facet B_1;
- For $j = 2, \ldots, n - 2$, $\mathcal{Q}_j$ starts with B_{i_j} and continues with all the facets in $H \setminus (\mathcal{Q}_1 \cup \ldots \cup \mathcal{Q}_{j-1} \cup \{B_{i_j}\})$ which are type 2 with respect to the facets in $\mathcal{Q}_1 \cup \ldots \cup \mathcal{Q}_{j-1}$ and those which already precede them in $\mathcal{Q}_j$.

To prove that this is a shelling, we show that

$$\{B_1, B_2, B_3, \ldots, B_{i_{j+1}-1}\} \subseteq \mathcal{Q}_1 \cup \ldots \cup \mathcal{Q}_j \quad \text{for } j = 1, \ldots, n - 2, \qquad (7.3)$$

where we make $i_{n-1} = t + 1$. This holds trivially for $j = 1$ since $i_2 = 2$. Assume that $j > 1$ and (7.3) holds for $j - 1$. We have $B_{i_j} \in \mathcal{Q}_j$ by construction. On the other hand, $B_{i_j+1}, \ldots, B_{i_{j+1}-1}$ are type 2 in $B_1, \ldots, B_t$. Using the induction hypothesis, and proceeding step by step, it follows that all these facets must belong to $\mathcal{Q}_j$ (unless they already appeared before in some $\mathcal{Q}_i$ ($i < j$)). Therefore (7.3) holds.

Now we claim that the facets B_{i_j} are type 1 in the new sequence: if $j > 1$, one of the symbols of B_{i_j} makes its first appearance, and the other two (say p, q) are such that $pq \subseteq B_k$ for some $k < i_j$ (since $B_1, \ldots, B_t$ is a shelling), and then we apply (7.3) to get $pq \subseteq C$ for some $C \in \mathcal{Q}_1 \cup \ldots \cup \mathcal{Q}_{j-1}$. Of course, all the others facets are type 2 by construction in the new sequence, hence our new enumeration is indeed a shelling.

For $j = 1, \ldots, n - 2$, let $\Omega_j = (W_j, E_j)$ be the graph with vertex set $W_j = B_1 \cup \ldots \cup B_{i_j}$ and edges $p \text{ --- } q$ whenever $pq \subseteq B$ for some $B \in \mathcal{Q}_1 \cup \ldots \cup \mathcal{Q}_j$. We say that a vertex $p \in V_i$ has *color* i. We claim that

$$\text{if } pq, qr \in E_j \text{ and } pqr \in H, \text{ then } pr \in E_j. \qquad (7.4)$$

Indeed, if $pq, qr \in E_j$, then $pqx, qry \in \mathcal{Q}_1 \cup \ldots \cup \mathcal{Q}_j$ for some $x, y \in V$. Since $pqr \in H$, it follows that $pqr \in \mathcal{Q}_1 \cup \ldots \cup \mathcal{Q}_j$ and so $pr \in E_j$.

On the other hand, if we try to construct Ω_j from Ω_{j-1}, we have to adjoin a new vertex corresponding to the letter p making its first appearance in $B_{i_j} = pqr$, and two new edges $p \text{ --- } q$ and $p \text{ --- } r$. The effect on the graph of adjoining a facet

$B = xyz \in Q_j \setminus \{B_{i_j}\}$ (if any), is that of adjoining an edge y — z in the presence of two edges x — y and x — z involving at most two colors. Of course, $xyz \in H$ if and only if x, y, z have at most two colors, since $H = H^1(\Gamma)$ and we assume all the connected components of Γ to be complete.

In view of (7.4), we define a graph $\overline{\Omega}_j$ having as vertices the monochromatic connected components of Ω_j (i.e. maximal sets of vertices of the same color which induce a connected subgraph of Ω_j), and having an edge X — Y between two distinct vertices X and Y if Ω_j has an edge x — y with $x \in X$ and $y \in Y$. By (7.4), this is equivalent to saying that Ω_j has an edge x — y for all $x \in X$ and $y \in Y$. Note also that if X — Y is an edge of $\overline{\Omega}_j$, then X and Y have different colors.

Next we prove that:

$$\overline{\Omega}_j \text{ is a tree for } j = 1, \ldots, n - 2. \tag{7.5}$$

This is obvious for $j = 1$, hence assume that $j > 1$ and $\overline{\Omega}_{j-1}$ is a tree. If we construct Ω_j from Ω_{j-1}, we have to adjoin a new vertex corresponding to the letter p making its first appearance in $B_{i_j} = pqr$, and two new edges p — q and p — r. Furthermore, by (7.4) and the comments following it, Ω_j is obtained by successively adjoining new edges x — z whenever x — y and x — z are already edges with $xyz \in H$.

Assume that $p \in V_d, q \in V_{d'}$ and $r \in V_{d''}$. Let D (respectively D', D'') denote the monochromatic connected component of p (respectively q, r) in Ω_{j-1}.

If $d = d'$, then the new edges of Ω_j connect p to every vertex in D and in every monochromatic connected component adjacent to D in $\overline{\Omega}_{j-1}$, hence $\overline{\Omega}_j = \overline{\Omega}_{j-1}$ and is therefore a tree. Hence, by symmetry, we may assume that $d \neq d', d''$. Since $pqr \in H = H^1(\Gamma)$, it follows that $d' = d''$ and so $D' = D''$ (since there exists an edge q — r in Ω_{j-1} due to pqr being type 1). Now, if there exists in $\overline{\Omega}_{j-1}$ a monochromatic connected component F of color d adjacent to D', it is easy to see that the new edges of Ω_j connect p to every vertex in F and in every monochromatic connected component adjacent to F in $\overline{\Omega}_{j-1}$, so we get once again $\overline{\Omega}_j = \overline{\Omega}_{j-1}$. Hence we may assume that in $\overline{\Omega}_{j-1}$ there exists no monochromatic connected component of color d adjacent to D'. It follows that the new edges of Ω_j connect p to every vertex in D'. Hence p is going to constitute a new monochromatic connected component of its own in Ω_j and so $\overline{\Omega}_j$ can be obtained from $\overline{\Omega}_{j-1}$ by adjoining the new vertex $\{p\}$ and the new edge $\{p\}$ — D'. Therefore $\overline{\Omega}_j$ is a tree and so (7.5) holds for every j by induction.

In particular, $\overline{\Omega}_{n-2}$ is a tree. However, since $P_{\leq 2}(V) \subseteq H$ and $\mathcal{H}$ is pure, every 2-subset of V eventually occurs as a subset of some $\mathcal{B}_i$ in the sequence, hence Ω_{n-2} is the complete graph with vertex set V, and so $\overline{\Omega}_{n-2}$ should be the complete graph with vertex set $\{V_1, \ldots, V_m\}$, which is not a tree since $m > 2$. We reached thus a contradiction, so we can deduce that $\mathcal{H}$ is not shellable for the particular case of H considered.

Finally, we consider the general case. Let Γ' be the graph obtained from Γ by adding all possible edges to each connected component and let $H' = H^1(\Gamma')$. Since $\mathcal{H}$ is boolean representable, we have $H \subseteq H^1(\Gamma) \subseteq H'$ by Lemma 6.4.4. Suppose that $B_1, \ldots, B_t$ is a shelling of $\mathcal{H}$. Since all the facets of $\mathcal{H}$ have dimension 2 by Lemma 7.2.4, it follows that $B_1, \ldots, B_t$ are facets of (V, H') and every $X \in P_{\leq 2}(V)$ occurs as a subset of some B_i. Therefore, if we extend the sequence $B_1, \ldots, B_t$ by adjoining the remaining facets of (V, H'), we obtain a shelling of (V, H'), a contradiction in view of our discussion of the particular case. Therefore $\mathcal{H}$ is not shellable. $\square$

We can now obtain:

Theorem 7.2.8. *Let* $\mathcal{H} \in \mathrm{BPav}(2)$. *Then the following conditions are equivalent:*

(i) $\mathcal{H}$ *is shellable;*
(ii) $\lceil Fl \rceil \, \mathcal{H}$ *contains at most two connected components or contains exactly one nontrivial connected component.*

Proof. Write $\lceil Fl \rceil \, \mathcal{H} \in \mathcal{C}_u^v$. Then $u \geq 1$ by Theorem 6.4.9. Now we combine Lemmas 7.2.1, 7.2.5 and 7.2.7. $\square$

We can use Theorem 7.2.8 to produce a characterization of the complexes in BPav(2) which are sequentially Cohen-Macaulay. For details, we shall refer the reader to [7, 19, 51].

Let $\mathcal{H} = (V, H)$ be a trim simplicial complex of dimension d. For $m = 0, \ldots, d$, we define the complex $\mathrm{pure}_m(\mathcal{H}) = (V_m, H_m)$, where

$$V_m = \cup(H \cap P_{m+1}(V)), \quad H_m = \cup_{X \in H \cap P_{m+1}(V)} 2^X.$$

It is easy to check that $\mathrm{pure}_m(\mathcal{H})$ is a trim pure complex of dimension m.

In view of [19, Theorem 3.3], we say that $\mathcal{H}$ is *sequentially Cohen-Macaulay* if

$$\tilde{H}_k(\mathrm{pure}_m(\mathrm{lk}(X))) = 0$$

for all $X \in H$ and $k < m \leq d$, where $\tilde{H}_k$ denotes the kth reduced homology group.

Since we are only considering dimension 2 here, we only need to deal with low dimensions. We proceed next do define $\tilde{H}_k$ for $k \geq 1$ since it is enough to use $\tilde{H}_1$ in our proof.

Let $\mathcal{J} = (V, J)$ be a simplicial complex. Fix a total ordering of V and let $k \geq 1$. Let $C_k(\mathcal{J})$ denote all the formal sums of the form $\sum_{i \in I} n_i X_i$ with $n_i \in \mathbb{Z}$ and $X_i \in J \cap P_{k+1}(V)$ (distinct). Given $X \in J \cap P_{k+1}(V)$, write $X = x_0 x_1 \ldots x_k$ with $x_0 < \ldots < x_k$. We define

$$X \partial_k = \sum_{i=0}^{k} (-1)^i (X \setminus \{x_i\}) \in C_{k-1}(\mathcal{J})$$

and extend this by linearity to a homomorphism $\partial_k : C_k(\mathcal{J}) \to C_{k-1}(\mathcal{J})$. Then the *kth reduced homology group* of $\mathcal{J}$ (which coincides with the kth homology group since $k \geq 1$) is defined as the quotient

$$\tilde{H}_k(\mathcal{J}) = \text{Ker}\, \partial_k / \text{Im}\, \partial_{k+1}.$$

We can now characterize the sequentially Cohen-Macaulay complexes in BPav(2).

Corollary 7.2.9. *Let* $\mathcal{H} \in$ BPav(2)*. Then the following conditions are equivalent:*

(i) $\mathcal{H}$ *is sequentially Cohen-Macaulay;*
(ii) $\mathcal{H}$ *is shellable;*
(iii) ΓFl $\mathcal{H}$ *contains at most two connected components or contains exactly one nontrivial connected component.*

Proof. It is known [7, 51] that every shellable simplicial complex is sequentially Cohen-Macaulay. In view of Theorem 7.2.8, it remains to be shown that $\mathcal{H}$ is not sequentially Cohen-Macaulay whenever $\Gamma = \Gamma$Fl $\mathcal{H} \in \mathcal{C}_m^n$ with $m \geq 2$ and $n \geq 3$.

Let $\mathcal{H} = (V, H)$ be such a complex. Let A_1, A_2 and A_3 denote three distinct connected components of $\mathcal{H}$ with A_1, A_2 nontrivial. Fix vertices $a_i \in A_i$ for $i = 1, 2, 3$. We may assume without loss of generality that $a_1 < a_2 < a_3$. Let $X = \emptyset$ and $\mathcal{J} = \text{pure}_2(\text{lk}(X)) = \text{pure}_2(\mathcal{H})$. Let $a_i \longrightarrow b_i$ be an edge of Γ for $i = 1, 2$. Then $a_1 b_1 a_2, a_1 b_1 a_3, a_2 b_2 a_3 \in H$ and so $a_1 a_2, a_1 a_3, a_2 a_3$ are faces of $\mathcal{J}$. Let $u = a_1 a_2 - a_1 a_3 + a_2 a_3 \in C_2(\mathcal{J})$. Since

$$u\partial_1 = (a_1 a_2 - a_1 a_3 + a_2 a_3)\partial_1 = (a_2 - a_1) - (a_3 - a_1) + (a_3 - a_2) = 0,$$

we have $u \in \text{Ker}\, \partial_1$. Note that u corresponds to the boundary of the triangle $a_1 a_2 a_3$.

Let Y contain all 2-subsets of V intersecting two distinct connected components of Γ. Given $w = \sum_{i \in I} n_i X_i \in C_2(\mathcal{J})$, let $w\zeta$ denote the sum of the n_i such that $X_i \in Y$.

Since $\mathcal{H}$ is boolean representable, it follows from Lemma 6.4.4 that every face pqr in $\mathcal{H}$ (and therefore in $\mathcal{J}$) has at least two vertices in the same connected component, hence $(pqr)\partial_2\zeta$ is even. It follows that $(\text{Im}\, \partial_2)\zeta \subseteq 2\mathbb{Z}$. Since $u\zeta = 1$, then $\text{Im}\, \partial_2 \subset \text{Ker}\, \partial_1$ and so $\tilde{H}_1(\mathcal{J}) \neq 0$. Therefore $\mathcal{H}$ is not sequentially Cohen-Macaulay as required.

In other words, $a_1 a_2 a_3$ is not a face, hence u is a cycle which is not a boundary, yielding the desired nontrivial homology. $\square$

7.3   Shellability Within Pav(2)

It is not likely to obtain a generalization of Theorem 7.2.8 to Pav(2) because a simplicial complex is not boolean representable precisely when its flats are not rich enough to represent it!

However, we can discuss, for a given graph in C_t^m ($t \geq 2, m \geq 3$) the possibility of making it the graph of flats of some shellable $\mathcal{H} \in \mathrm{Pav}(2)$. It can be done in most cases:

Theorem 7.3.1. *Let* $\Gamma = (V, E) \in C_t^m$ *with* $t \geq 2, m \geq 3$. *Then the following conditions are equivalent:*

(i) $\Gamma \cong \Gamma\mathrm{Fl}(V, H)$ *for some shellable* $(V, H) \in \mathrm{Pav}(2)$;
(ii) Γ *is not of the form* $K_r \sqcup K_s \sqcup K_1$ *for some* $r, s > 1$.

Proof. (i) $\Rightarrow$ (ii). Assume that $\Gamma = \Gamma\mathrm{Fl}(V, H)$ for some $(V, H) \in \mathrm{Pav}(2)$. Suppose that $\Gamma \cong K_r \sqcup K_s \sqcup K_1$ for some $r, s > 1$, corresponding to connected components V_1, V_2, V_3, respectively. Suppose that $X \in H$ intersects all three connected components. Write $X = x_1 x_2 x_3$ with $x_i \in V_i$.

Let $y \in V \setminus \{x_1, x_2\}$. Then either $y = x_3$ or $y \in \mathrm{nbh}(x_1) \cup \mathrm{nbh}(x_2)$. Note that, in the latter case we must have $yx_1 x_2 \in H$ by Lemma 6.4.3. Thus $yx_1 x_2 \in H$ in any case and so $x_1 x_2 \in \mathrm{Fl}(V, H)$. It follows that x_1 is adjacent to x_2, a contradiction.

Therefore no element of H intersects all three connected components, hence $H \subseteq H^1(\Gamma)$ and so (V, H) is boolean representable by Lemma 6.4.4. Thus (V, H) is not shellable by Lemma 7.2.7.

(ii) $\Rightarrow$ (i). Let $V = V_1 \cup \ldots \cup V_m$ be the decomposition of V in its connected components, where $V_1, \ldots, V_t$ are the nontrivial ones. For each $i \in \{1, \ldots, m\}$, fix $a_i \in V_i$. For each $i \in \{1, \ldots, t\}$, fix also $b_i \in V_i \setminus \{a_i\}$. We may assume that:

(AS1)   If there exist non complete components, the first component is among them;

(AS2)   If the ith component is not complete, then a_i and b_i are not adjacent.

We define

$$H = H^1(\Gamma) \cup \{a_1 a_i a_j \mid 1 < i < j \leq t\}$$
$$\cup \{a_1 a_i a_j \mid 1 < i \leq t < j < m\}$$
$$\cup \{b_1 b_i a_m \mid 1 < i \leq t < m\}.$$

Clearly, $(V, H) \in \mathrm{Pav}(2)$. Note also that, given three distinct connected components, there is at most one face of H intersecting them all. We claim that

$$\mathrm{Fl}(V, H) = P_{\leq 1}(V) \cup E \cup \{V\}. \tag{7.6}$$

Since $H^1(\Gamma) \subseteq H$, it is easy to see that $P_{\leq 1}(V) \cup E \cup \{V\} \subseteq \mathrm{Fl}(V, H)$.

To prove the direct inclusion, we start by the following remarks:

$$\text{if } |X| \geq 3 \text{ and } X \text{ is not an anticlique, then } \overline{X} = V. \tag{7.7}$$

Indeed, if this happens then X contains some facet $Y \in H^1(\Gamma) \subseteq H$ and so $\overline{X} = V$ by Proposition 4.2.3.

We show next that

$$\text{if } X \text{ intersects 3 connected components, then } \overline{X} = V. \tag{7.8}$$

We may assume that $|X| = 3$. Suppose first that $X = pqr$ intersects two nontrivial connected components, say $p \in V_i$ and $q \in V_j$. Take $p' \in \text{nbh}(p)$ and $q' \in \text{nbh}(q)$. It is impossible to have $pq'r, p'qr \in H$ simultaneously, hence $p' \in \overline{X}$ or $q' \in \overline{X}$. In any case, we may apply (7.7) to get $\overline{X} = V$.

Thus we may assume that q and r are isolated points. Then $xqr \notin H$ for every $x \in V_1$ and so $V_1 \subseteq \overline{X}$ and so $\overline{X} = V$ by (7.7). Therefore (7.8) holds.

Finally, we claim that

$$\overline{pq} = V \text{ whenever } p, q \text{ are two non adjacent vertices.} \tag{7.9}$$

Assume first that $p, q \in V_i$. Then $V \setminus V_i \subseteq \overline{pq}$ since $pqx \notin H$ for every $x \in V \setminus V_i$. Thus $\overline{pq} = V$ by (7.8).

Thus we may assume that $p \in V_i$ and $q \in V_j$ with $i \neq j$. Suppose first that there exists some $k \in \{1, \ldots, t\} \setminus \{i, j\}$. Then $pqr \notin H$ for some $r \in V_k$, hence $r \in \overline{pq}$ and so $\overline{pq} = V$ by (7.8). Thus we may assume that $i = 1, j = 2$ and $t = 2$. If $m \geq 4$, then either $pqa_3 \notin H$ or $pqa_m \notin H$, hence we get a_3 or a_m into $\overline{pq}$ and so $\overline{pq} = V$ by (7.8). If $m = 3$, then we may assume that $p = b_1$ (otherwise $pqa_3 \notin H$ and we use (7.8) as before). It follows from (ii) and (AS1), (AS2) that $a_1 \notin \text{nbh}(p)$. Hence $a_1pq \notin H$ and so $a_1 \in \overline{pq}$. Now $a_1qa_3 \notin H$ yields $a_3 \in \overline{V}$ and so (7.9) follows from (7.8).

It follows that (7.6) holds and so $\Gamma\text{Fl}(V, H) = \Gamma$. All we need now is to prove shellability. Let $\mathcal{B}$ denote the set of facets of dimension 2 of (V, H). For $i \in \{1, \ldots, t\}$, let $\mathcal{B}_i = \mathcal{B} \cap 2^{V_i}$. For $1 \leq i < j \leq m$, let

$$\mathcal{B}_{ij} = (\mathcal{B} \cap 2^{V_i \cup V_j}) \setminus (\mathcal{B}_i \cup \mathcal{B}_j).$$

It is easy to check that

$$\begin{aligned}
\mathcal{B} = {} & \mathcal{B}_1 \cup (\cup_{i=2}^t (\mathcal{B}_{1i} \cup \mathcal{B}_i)) \cup (\cup_{1<i<j\leq t}(\{a_1a_ia_j\} \cup \mathcal{B}_{ij})) \\
& \cup (\cup_{t<k<m}\mathcal{B}_{1k} \cup (\cup_{i=2}^t(\{a_1a_ia_k\} \cup \mathcal{B}_{ik}))) \\
& \cup \mathcal{B}_{1m} \cup (\cup_{i=2}^t(\{b_1b_ia_m\} \cup \mathcal{B}_{im})),
\end{aligned}$$

where the two last lines are omitted if $t = m$.

We claim that we can use this decomposition, followed by arbitrary enumeration of the facets of dimension 1, to produce a shelling. For instance, for $\mathcal{B}_1$ we write $V_1 = c_{11}c_{12}\ldots c_{in_1}$ with $c_{11} = a_1$, $c_{12} \in \text{nbh}(a_1)$, and $c_{1j} \in \text{nbh}(c_{1j'})$ for some $j' < i$. Then we enumerate successively:

- Facets of the form $c_{11}c_{12}c_{1j}$, for $j = 3, \ldots, n_1$;
- For each $i = 3, \ldots, n_1$: all the remaining facets of the form $c_{1i}c_{1i'}c_{1j}$;
- All the remaining facets in $\mathcal{B}_1$.

We adapt this same technique to deal with each one of the segments $\mathcal{B}_i$ and $\mathcal{B}_{ij}$. It is easy to check directly that the other facets also satisfy (7.1), hence we have a shelling as claimed. $\square$

7.4   Betti Numbers

We compute next the Betti numbers for the particular case of shellable $\mathcal{H} \in \text{Pav}(2)$. We define

$$\text{Sing } \mathcal{H} = \{p \in V \mid pqr \notin H \text{ for all } q, r \in V \text{ distinct}\}.$$

Denote by $\text{fct}_i \mathcal{H}$ the set of all facets of dimension i in $\mathcal{H}$.

Theorem 7.4.1. *Let $\mathcal{H} = (V, H) \in \text{Pav}(2)$ be shellable. Then:*

(i) $w_1(\mathcal{H}) = |\text{fct}_1 \mathcal{H}| - |\text{Sing } \mathcal{H}|$;
(ii) $w_2(\mathcal{H}) = |\text{fct } \mathcal{H}| - |\text{Sing } \mathcal{H}| + \frac{|V|(3-|V|)}{2} - 1.$

Proof. We adapt the construction of the graph sequence in the proof of Lemma 7.2.7. Let $m = |\text{fct}_2 \mathcal{H}|$ and $t = |\text{fct}_1 \mathcal{H}|$. Let $B_1, \ldots, B_{m+t}$ be a shelling of $\mathcal{H}$. By Lemma 7.1.4, we may assume that $|B_i| = 3$ for $i \leq m$ and $|B_i| = 2$ if $m < i \leq m + t$.

For $i = 1, \ldots, m$, let $\Omega_i = (V_i, E_i)$ be the graph with vertex set $V_i = B_1 \cup \ldots \cup B_i$ and edges $p - q$ whenever $pq \subseteq B$ for some $B \in B_1 \cup \ldots \cup B_i$. Let h_i denote the number of homology facets among $B_1, \ldots, B_i$. We claim that

$$h_i = i - |E_i| + |V_i| - 1 \tag{7.10}$$

for $i = 1, \ldots, m$. Indeed, since $\dim \mathcal{H} = 2$, we have $m \geq 1$ and $1 - |E_1| + |V_1| - 1 = 1 - 3 + 3 - 1 = 0 = h_1$. Hence (7.10) holds for $i = 1$.

Assume now that $i \in \{2, \ldots, m\}$ and (7.10) holds for $i - 1$. We consider three cases:

Case 1: $B_i \not\subseteq V_{i-1}$.
   Then $|V_i| = |V_{i-1}| + 1$ and $|E_i| = |E_{i-1}| + 2$. Hence

$$h_i = h_{i-1} = i - 1 - |E_{i-1}| + |V_{i-1}| - 1$$
$$= i - 1 - |E_i| + 2 + |V_i| - 1 - 1 = i - |E_i| + |V_i| - 1.$$

Case 2: $B_i \subseteq V_{i-1}$ and B_i is not a homology facet.
Then $|V_i| = |V_{i-1}|$ and $|E_i| = |E_{i-1}| + 1$. Hence

$$h_i = h_{i-1} = i - 1 - |E_{i-1}| + |V_{i-1}| - 1$$
$$= i - 1 - |E_i| + 1 + |V_i| - 1 = i - |E_i| + |V_i| - 1.$$

Case 3: $B_i \subseteq V_{i-1}$ and B_i is a homology facet.
Then $|V_i| = |V_{i-1}|$ and $|E_i| = |E_{i-1}|$. Hence

$$h_i = h_{i-1} + 1 = i - 1 - |E_{i-1}| + |V_{i-1}| - 1 + 1$$
$$= i - 1 - |E_i| + |V_i| = i - |E_i| + |V_i| - 1.$$

Thus (7.10) holds for i in all three cases. By induction, it follows that (7.10) holds for m and so by Theorem 7.1.1

$$w_2(\mathcal{H}) = h_m = m - |E_m| + |V_m| - 1. \tag{7.11}$$

Now V_m consists of all the points that appear in some $pqr \in H$, hence $|V_m| = |V| - |\text{Sing } \mathcal{H}|$. On the other hand, since $\mathcal{H}$ is simple, for all distinct $p, q \in V$, either $pq \in E_m$ or $pq \in \text{fct}_1 \mathcal{H}$. Hence $|E_m| = \binom{|V|}{2} - |\text{fct}_1 \mathcal{H}|$ and so (7.11) yields

$$w_2(\mathcal{H}) = m - |E_m| + |V_m| - 1 = m - \binom{|V|}{2} + t + |V| - |\text{Sing } \mathcal{H}| - 1$$

$$= |\text{fct } \mathcal{H}| - |\text{Sing } \mathcal{H}| + \frac{|V|(3 - |V|)}{2} - 1$$

and (ii) holds.

Finally, since there exist $|\text{Sing } \mathcal{H}|$ points which do not appear in the facets of dimension 2, they will make their first appearance in the facets of dimension 1. This ensures that at least $|\text{Sing } \mathcal{H}|$ facets of dimension 1 are not homology facets. It is immediate that all the other facets of dimension 1 must be homology facets, since all their points have already appeared before. Thus $w_1(\mathcal{H}) = |\text{fct}_1 \mathcal{H}| - |\text{Sing } \mathcal{H}|$ by Theorem 7.1.1. □

Before discussing the boolean representable case, we prove the following lemma:

Lemma 7.4.2. *Let* $\mathcal{H} = (V, H) \in \text{BPav}(2)$. *Then* $\text{fct}_1 \mathcal{H} = P_2(V')$ *for*

$$V' = \{p \in V \mid nbh(p) = \emptyset \text{ in } \Gamma\text{Fl } \mathcal{H}\}.$$

Proof. Write $\Gamma = \overline{\Gamma\text{Fl}} \ \mathcal{H}$. Let $ab \in \text{fct}_1 \ \mathcal{H}$. Suppose that $nbh(a) \neq \emptyset$. If $b \in nbh(a)$, then $\overline{ab} \subset V$ and $abx \in H$ for any $x \in V \setminus \overline{ab}$. contradicting $ab \in \text{fct}_1 \ \mathcal{H}$. On the other hand, if $b \notin nbh(a)$, we take $c \in nbh(a)$ to get $abc \in H^0(\Gamma) \subseteq H$ by Lemma 6.4.3, a contradiction too. Thus $nbh(a) = \emptyset = nbh(b)$ and so $\text{fct}_1 \ \mathcal{H} \subseteq P_2(V')$.

Conversely, let $ab \in P_2(V')$. Suppose that $abc \in H$ for some $c \in V \setminus ab$. By Lemma 6.4.4, we have $abc \in H^1(\Gamma)$, contradicting $ab \subseteq V'$. Thus $ab \in \text{fct}_1 \, \mathcal{H}$ as required. $\square$

Theorem 7.4.3. *Let* $\mathcal{H} = (V, H) \in \text{BPav}(2)$ *be shellable with* $\Gamma\text{Fl} \, \mathcal{H} \in \mathcal{C}_k^n$. *Then:*

(i) $w_1(\mathcal{H}) = |\text{fct}_1 \, \mathcal{H}| = \binom{n-k}{2}$;
(ii) $w_2(\mathcal{H}) = |\text{fct} \, \mathcal{H}| + \frac{|V|(3-|V|)}{2} - 1$.

Proof. In view of Theorem 7.4.1 and Lemma 7.4.2, it suffices to show that Sing $\mathcal{H} = \emptyset$. Let $p \in V$. By Theorem 6.4.9, there exists some $q \in V \setminus \{p\}$ with $\text{nbh}(q) \neq \emptyset$ in $\Gamma\text{Fl} \, \mathcal{H}$. By Lemma 7.4.2, pq is not a facet of $\mathcal{H}$ and so $p \notin \text{Sing} \, \mathcal{H}$. $\square$

In view of Theorem 7.1.1, the combination of Theorems 7.2.8 and 7.4.3 allows an efficient determination of the homotopy type of $\| \, \mathcal{H} \, \|$ when $\mathcal{H} \in \text{BPav}(2)$ is shellable. In particular, we obtain the following corollary.

Corollary 7.4.4. *Let* $\mathcal{H} \in \text{BPav}(2)$ *be such that* $\Gamma\text{Fl} \, \mathcal{H}$ *contains at most two connected components or contains exactly one nontrivial connected component. Then* $\mathcal{H}$ *has the homotopy type of a wedge of spheres of computable number and dimension.*

This is in contrast with the situation for arbitrary simplicial complexes of dimension 2, where the homotopy type is undecidable, in view of the following argument. By [46, Theorem 7.45], every finitely presented group G occurs as the fundamental group of a simplicial complex $\mathcal{H} \, (G)$ of dimension 2. Now $\mathcal{H} \, (G)$ is *simply connected* (i.e. has the homotopy type of a point) if and only if G is trivial. Since it is undecidable whether or not a finitely presented group is trivial [37, Theorem IV.4.1], it is undecidable whether or not $\mathcal{H}(G)$ is simply connected.

Chapter 8
Operations on Simplicial Complexes

We consider in this chapter various natural operations on simplicial complexes and study how they relate to boolean representability. The particular case of restrictions will lead us to introduce prevarieties of simplicial complexes and finite basis problems.

8.1 Boolean Operations

Boolean representability behaves badly with respect to intersection and union, even in the paving case, as we show next.

Given simplicial complexes $\mathcal{H} = (V, H)$ and $\mathcal{H}' = (V, H')$, we define

$$\mathcal{H} \cap \mathcal{H}' = (V, H \cap H') \quad \text{and} \quad \mathcal{H} \cup \mathcal{H}' = (V, H \cup H').$$

First, we recall a well-known fact.

Proposition 8.1.1. *Every simplicial complex* $\mathcal{H} = (V, H)$ *is the intersection of matroids on* V.

Proof. For every circuit X of $\mathcal{H}$, let $M_X = (V, H_X)$, where H_X consists of all subsets of V not containing X. We claim that M_X is a matroid.

Clearly, H_X is closed under taking subsets, so let $I, J \in H_X$ be such that $|I| = |J| + 1$. We may assume that $J \not\subseteq I$. If $|J \cap X| < |X| - 1$, then $J \cup \{p\} \in H_X$ for every $p \in V$, hence we may assume that $|J \cap X| = |X| - 1$, so $X \setminus J = \{x\}$ for some $x \in X$. Since $J \not\subseteq I$, there exists some $i \in I \setminus (J \cup \{x\})$. Then $J \cup \{i\} \in H_X$ and so M_X is a matroid.

© Springer International Publishing Switzerland 2015
J. Rhodes, P.V. Silva, *Boolean Representations of Simplicial Complexes
and Matroids*, Springer Monographs in Mathematics,
DOI 10.1007/978-3-319-15114-4_8

To complete the proof, it suffices to show that

$$H = \bigcap_{X \in C} H_X, \tag{8.1}$$

where C denotes the set of all circuits of $\mathcal{H}$. The direct inclusion is obvious. For the opposite, let $Y \in 2^V \setminus H$. Then Y contains some $X \in C$, hence $Y \notin H_X$ and (8.1) holds as required. $\square$

In view of Theorem 5.2.10, we immediately get:

Corollary 8.1.2. *Every simplicial complex is the intersection of boolean representable simplicial complexes.*

Thus, like matroids, boolean representable simplicial complexes are not closed under intersection.

The next counterexample shows that paving does not help:

Example 8.1.3. Let $V = \{1, \ldots, 4\}$, $H = P_{\leq 2}(V) \cup \{123, 124\}$ and $H' = P_{\leq 2}(V) \cup \{123, 134\}$. Then $(V, H), (V, H') \in \mathrm{BPav}(2)$ but $(V, H \cap H')$ is not boolean representable.

Indeed, both (V, H) and (V, H') are T_2 and therefore boolean representable (see Example 5.2.11). Since $H \cap H' = P_{\leq 2}(V) \cup \{123\}$, then $(V, H \cap H')$ is T_1 and therefore not boolean representable (also by Example 5.2.11).

Union does not behave any better.

Example 8.1.4. Let $V = \{1, \ldots, 6\}$, $J_1 = P_{\leq 3}(V) \setminus \{123, 125, 135, 235, 146, 246, 346, 456\}$ and $J_2 = P_{\leq 2}(V) \cup \{123, 124, 125, 126\}$. Then $(V, J_1), (V, J_2) \in \mathrm{BPav}(2)$, but $(V, J_1 \cup J_2)$ is not boolean representable.

Indeed, it is easy to check that $1235 \in \mathrm{Fl}(V, J_1)$. Since $|xyz \cap 1235| = 2$ for every $xyz \in J_1$, it follows from Proposition 6.1.2 that (V, J_1) is boolean representable. Similarly, since $12 \in \mathrm{Fl}(V, J_2)$, it follows that (V, J_2) is boolean representable.

Now $J_1 \cup J_2 = P_{\leq 3}(V) \setminus \{135, 235, 146, 246, 346, 456\}$ and it is straightforward to check that in this simplicial complex $\overline{13} = \overline{14} = \overline{34} = V$. By Proposition 6.1.2, $(V, J_1 \cup J_2)$ is not boolean representable.

However, closure under union can be satisfied in some circumstances.

Proposition 8.1.5. *Let $(V, H_1), (V, H_2) \in \mathrm{BPav}(2)$ be such that*

$$X \in \mathrm{Fl}(V, H_i) \setminus \{V\} \Rightarrow |X| \leq 3 \tag{8.2}$$

holds for $i = 1, 2$. Then $(V, H_1 \cup H_2) \in \mathrm{BPav}(2)$.

Proof. Let $\mathrm{Cl}_1, \mathrm{Cl}_2$ and Cl denote respectively the closure operators of $(V, H_1), (V, H_2)$ and $(V, H_1 \cup H_2)$.

Let $abc \in H_1 \cup H_2$. We may assume that $abc \in H_1$. Since (V, H_1) is boolean representable, we may assume by Proposition 6.1.2 that $c \notin \mathrm{Cl}_1(ab)$. If

$Cl_1(ab) = ab$, then $abx \in H_1$ for every $x \in V \setminus ab$, hence $c \notin ab = Cl(ab)$. Hence we may assume that $ab \subset Cl_1(ab)$. By (8.2), we may write $Cl_1(ab) = abd$ for some $d \in V \setminus abc$.

If $abd \in H_2$, then $abx \in H_1 \cup H_2$ for every $x \in V \setminus ab$ and so $Cl(ab) = ab$. If $abd \notin H_2$, then it is easy to check that $Cl(ab) = abd$. Thus, we get $c \notin Cl(ab)$ in any case and so $(V, H_1 \cup H_2)$ is boolean representable by Proposition 6.1.2. $\square$

8.2 Truncation

Given a simplicial complex $\mathcal{H} = (V, H)$ and $k \geq 0$, the k-*truncation* of $\mathcal{H}$ is the simplicial complex $\mathcal{H}_k = (V, H_k)$ defined by $H_k = H \cap P_{\leq k}(V)$. In line with the negative results from Sect. 8.1, we show that boolean representability is not preserved under truncation. However, we succeed in characterizing those simplicial complexes which are truncations of boolean representable simplicial complexes.

The next example shows that boolean representability is not preserved under truncation, even in the simple case.

Example 8.2.1. Let $V = \{1, \ldots, 6\}$,

$$H = (P_{\leq 3}(V) \setminus \{135, 235, 146, 246, 346, 456\}) \cup \{1234, 1236, 1245, 1256\}$$

and $\mathcal{H} = (V, H)$. Then $\mathcal{H}$ is boolean representable, but $\mathcal{H}_3$ is not.

Indeed, it is easy to check that $P_{\leq 1}(V) \cup \{12, 1235\} \subseteq \mathrm{Fl}\, \mathcal{H}$. By Corollary 5.2.7, to show that $\mathcal{H}$ is boolean representable it suffices to show that every $X \in H$ admits an enumeration $x_1, \ldots, x_k$ satisfying (5.2). We may of course assume that $|X| > 2$. Hence X cannot contain both 4 and 6. Since 1235 is closed, it can be used to exclude 4 or 6, if one of them belongs to X. Hence we may assume that $X \subseteq 1235$. Since we only need to care about X being a 3-set, we are reduced to the cases $X \in \{123, 125\}$. Now $1 \subset 12 \subset 1235$ yields the desired chain of flats, and so $\mathcal{H}$ is boolean representable.

On the other hand, $\mathcal{H}_3$ is the simplicial complex $(V, J_1 \cup J_2)$ of Example 8.1.4, already proved not to be boolean representable.

Recall the notation fct $\mathcal{H}$ introduced in Sect. 4.1. In the following result, we characterize the flats of a truncation.

Proposition 8.2.2. *Let $\mathcal{H} = (V, H)$ be a simplicial complex and let $k \geq 0$. Then*

$$\mathrm{Fl}\, \mathcal{H}_k = \{X \in \mathrm{Fl}\, \mathcal{H} \mid \mathrm{fct}\, \mathcal{H}_k \cap 2^X = \emptyset\} \cup \{V\}.$$

Proof. Let $X \in \mathrm{Fl}\, \mathcal{H}_k \setminus \{V\}$. By Proposition 4.2.3, X cannot contain a facet of $\mathcal{H}_k$. Let $I \in H \cap 2^X$ and $p \in V \setminus X$. Since $I \notin \mathrm{fct}\, \mathcal{H}_k$, we have $|I| < k$ and so $I \in H_k$. Now $X \in \mathrm{Fl}\, \mathcal{H}_k$ yields $I \cup \{p\} \in H_k \subseteq H$. Therefore $X \in \mathrm{Fl}\, \mathcal{H}$ and the direct inclusion holds.

Conversely, assume that $X \in \mathrm{Fl}\, \mathcal{H}$ and X does not contain a facet of $\mathcal{H}_k$. Let $I \in H_k \cap 2^X$ and $p \in V \setminus X$. Since $H_k \subseteq H$ and $X \in \mathrm{Fl}\, \mathcal{H}$, we get $I \cup \{p\} \in H$. But I is not a facet of $\mathcal{H}_k$, hence $|I| < k$ and so $I \cup \{p\} \in H_k$. Thus $X \in \mathrm{Fl}\, \mathcal{H}_k$ as required. $\square$

Given a simplicial complex $\mathcal{H} = (V, H)$ of dimension d, we define

$$T(H) = \{T \subseteq V \mid \forall X \in H_d \cap 2^T \ \forall p \in V \setminus T \quad X \cup \{p\} \in H\}.$$

The following lemma is clear from the definition.

Lemma 8.2.3. *Let $\mathcal{H} = (V, H)$ be a simplicial complex. Then:*

(i) $T(H)$ is closed under intersection;
(ii) $\mathrm{Fl}\, \mathcal{H} \subseteq T(H)$.

Given $X \subseteq V$, we write $X \in H^T$ if X is a transversal of the successive differences for some chain of $T(H)$, that is, if there exists an enumeration $x_1, \ldots, x_k$ of X and $T_0, \ldots, T_k \in T(H)$ such that $T_0 \supset T_1 \supset \ldots \supset T_k$ and $x_i \in T_{i-1} \setminus T_i$ for $i = 1, \ldots, k$. Clearly, (V, H^T) is a simplicial complex.

Lemma 8.2.4. *Let $\mathcal{H} = (V, H)$ be a simplicial complex of dimension d. Then:*

(i) $(H^T)_{d+1} \subseteq H$;
(ii) $T(H) \subseteq \mathrm{Fl}(V, H^T)$;
(iii) (V, H^T) is boolean representable.

Proof. (i) We prove that

$$(H^T)_k \subseteq H \tag{8.3}$$

for $k = 0, \ldots, d + 1$ by induction on k.

The case $k = 0$ being trivial, assume that $k \in \{1, \ldots, d + 1\}$ and (8.3) holds for $k - 1$. Let $X \in (H^T)_k$. We may assume that $|X| = k$. Then there exists an enumeration $x_1, \ldots, x_k$ of X and $T_0, \ldots, T_k \in T(H)$ such that $T_0 \supset T_1 \supset \ldots \supset T_k$ and $x_i \in T_{i-1} \setminus T_i$ for $i = 1, \ldots, k$. Let $X' = \{x_2, \ldots, x_k\}$. Since $X' \in (H^T)_{k-1}$, it follows from the induction hypothesis that $X' \in H$. Now $|X'| \leq d$, $X' \subseteq T_1$ and $x_1 \in V \setminus T_1$, hence it follows from $T_1 \in T(H)$ that $X = X' \cup \{x_1\} \in H$. Thus (8.3) holds for $k = 0, \ldots, d + 1$.

(ii) Let $X \in T(H)$. Let $I \in H^T \cap 2^X$ and $p \in V \setminus X$. Since $I \in H^T$, there exists an enumeration $x_1, \ldots, x_k$ of I and $T_0, \ldots, T_k \in T(H)$ such that $T_0 \supset T_1 \supset \ldots \supset T_k$ and $x_i \in T_{i-1} \setminus T_i$ for $i = 1, \ldots, k$. Now by Lemma 8.2.3(i)

$$T_0 \cap X \supset T_1 \cap X \supset \ldots \supset T_k \cap X$$

is also a chain in $T(H)$ satisfying $x_i \in (T_{i-1} \cap X) \setminus (T_i \cap X)$ for $i = 1, \ldots, k$. Since $V \supset T_0 \cap X$ is also a chain in $T(H)$ and $p \in V \setminus (T_0 \cap X)$, we get $I \cup \{p\} \in H^T$ and so $X \in \mathrm{Fl}(V, H^T)$.

(iii) Let $X \in H^T$. Then X is a transversal of the partition of successive differences for some chain of $T(H)$. By (ii), this is also a chain in $\mathrm{Fl}(V, H^T)$. Now it follows easily from Corollary 5.2.7 that (V, H^T) is boolean representable. $\square$

Now we can prove the main result of this section:

Theorem 8.2.5. *Let $\mathcal{H} = (V, H)$ be a simplicial complex of dimension d. Then the following conditions are equivalent:*

(i) $H = J_{d+1}$ for some boolean representable simplicial complex (V, J);
(ii) $H = (H^T)_{d+1}$.

Furthermore, in this case we have $\mathrm{Fl}(V, H^T) = T(H)$.

Proof. (i) $\Rightarrow$ (ii). We start by showing that

$$\mathrm{Fl}(V, J) \subseteq T(H). \tag{8.4}$$

Let $F \in \mathrm{Fl}(V, J)$. Suppose that $X \in H_d \cap 2^F$ and $p \in V \setminus F$. Since $H \subseteq J$, it follows from $F \in \mathrm{Fl}(V, J)$ that $X \cup \{p\} \in J$. But now $|X| \le d$ implies $X \cup \{p\} \in J_{d+1} = H$ and so $F \in T(H)$. Therefore (8.4) holds.

Now let $X \in H$. Since $H \subseteq J$, it follows from Theorem 5.2.6 that there exists an enumeration $x_1, \ldots, x_k$ of X and $F_0, \ldots, F_k \in \mathrm{Fl}(V, J)$ such that $F_0 \supset F_1 \supset \ldots \supset F_k$ and $x_i \in F_{i-1} \setminus F_i$ for $i = 1, \ldots, k$. By (8.4), we have $F_0, \ldots, F_k \in T(H)$ and so $X \in H^T$. Since dim $\mathcal{H} = d$, then $X \in (H^T)_{d+1}$ and so $H \subseteq (H^T)_{d+1}$. Therefore $H = (H^T)_{d+1}$ by Lemma 8.2.4.

(ii) $\Rightarrow$ (i). This follows from Lemma 8.2.4(iii).

It remains to be proved that $\mathrm{Fl}(V, H^T) = T(H)$.

Let $X \in \mathrm{Fl}(V, H^T)$. Let $I \in H_d \cap 2^X$ and $p \in V \setminus X$. Then $I \in H^T$ by (ii) and so $X \in \mathrm{Fl}(V, H^T)$ yields $I \cup \{p\} \in H^T$. Since $|I| \le d$, we get $I \cup \{p\} \in (H^T)_{d+1} = H$ and so $X \in T(H)$. The opposite inclusion follows from Lemma 8.2.4(ii). $\square$

Example 8.2.6. The tetrahedron complex T_1 cannot be obtained as the truncation of a boolean representable simplicial complex.

Indeed, any boolean representable simplicial complex satisfies (PR) by Proposition 5.1.2, and it is easy to see that (PR) is preserved by truncation. However, T_1 does not satisfy (PR), as shown in Example 5.1.3.

Example 8.2.7. Let $V = \{1, \ldots, 6\}$ and $H = P_{\le 3}(V) \setminus \{135, 235, 146, 246, 346, 456\}$. Then (V, H) is not boolean representable, but (V, H^T) is and $H = H_3^T$.

Indeed, we have just remarked in Example 8.2.1 that (V, H) is not boolean representable. It is easy to check that $P_{\le 1}(V) \cup \{12, 1235, V\} \subseteq T(H)$. Considering the chains

$$V \supset 1235 \supset i \supset \emptyset, \quad V \supset 12 \supset j \supset \emptyset, \quad V \supset k \supset \emptyset$$

in $T(H)$ for $i \in \{1, 2, 3, 5\}$, $j \in \{1, 2\}$ and $k \in \{4, 6\}$, it is easy to see that $H \subseteq H^T$. In view of Lemma 8.2.4(i), it follows that $H = H_3^T$. Note that (V, H^T) is boolean representable by Lemma 8.2.4(iii).

8.3 Restrictions and Contractions

In the theory of matroids, restrictions and contractions are combined to build the concept of *minor*. As we shall see next, these two operators behave quite differently with respect to boolean representability.

Let $\mathcal{H} = (V, H)$ be a simplicial complex and $V' \subseteq V$ be nonempty. The *restriction* of $\mathcal{H}$ to V' is the simplicial complex $(V', H \cap 2^{V'})$.

The following result follows easily from the definitions:

Proposition 8.3.1. *The following classes of simplicial complexes are closed under restriction:*

  (i) *Boolean representable simplicial complexes;*
 (ii) *Paving simplicial complexes;*
(iii) *Graphic boolean simplicial complexes;*
 (iv) *Simplicial complexes satisfying (PR);*
  (v) *Matroids.*

Proof. (i) If M is an $R \times V$ boolean representation of $\mathcal{H} = (V, H)$, then $M[R, V']$ is a boolean representation of the restriction of $\mathcal{H}$ to V'.

(ii)–(v). Straightforward. $\square$

However, the next example shows that the restriction of a pure simplicial complex $\mathcal{H}$ needs not to be pure, even if $\mathcal{H} \in \mathrm{BPav}(2)$:

Example 8.3.2. Let $V = \{1, \ldots, 5\}$, $H = P_{\leq 3}(V) \setminus \{134, 234\}$ and $\mathcal{H} = (V, H)$. Then $\mathcal{H} \in \mathrm{BPav}(2)$ and is pure, but the restriction of $\mathcal{H}$ to $V' = \{1, \ldots, 4\}$ is not pure.

Indeed, it is immediate that $\mathcal{H} \in \mathrm{Pav}(2)$ and is pure. Since $12, 15, \ldots, 45 \in \mathrm{Fl}\,\mathcal{H}$, and every 3-set in H must contain one of these, $\mathcal{H}$ is boolean representable by Corollary 5.2.7.

However, the restriction $\mathcal{H}' = (V', P_{\leq 3}(V') \setminus \{134, 234\})$ is not pure since 34 is a facet of $\mathcal{H}'$.

The next result relates flats and closures in a simplicial complex and its restriction:

Proposition 8.3.3. *Let $\mathcal{H}' = (V', H')$ be a restriction of a simplicial complex $\mathcal{H} = (V, H)$. Then:*

 (i) *If $X \in \mathrm{Fl}\,\mathcal{H}$, then $X \cap V' \in \mathrm{Fl}\,\mathcal{H}'$;*
(ii) *$\mathrm{Cl}_{\mathcal{H}'}\, X \subseteq \mathrm{Cl}_{\mathcal{H}}\, X$ for every $X \subseteq V'$.*

Proof. (i) Let $I \in H' \cap 2^{X \cap V'}$ and $p \in V' \setminus (X \cap V')$, then $I \in H \cap 2^X$ and $p \in V \setminus X$, hence $I \cup \{p\} \in H$ since $X \in \mathrm{Fl}\, \mathcal{H}$. Thus $I \cup \{p\} \in H'$ and so $X \cap V' \in \mathrm{Fl}\, \mathcal{H}'$.

(ii) Let $X \subseteq V'$. By (i), we have $(\mathrm{Cl}_{\mathcal{H}} X) \cap V' \in \mathrm{Fl}\, \mathcal{H}'$. Since $X \subseteq (\mathrm{Cl}_{\mathcal{H}} X) \cap V'$, we get $\mathrm{Cl}_{\mathcal{H}'} X \subseteq (\mathrm{Cl}_{\mathcal{H}} X) \cap V'$. $\square$

We turn now our attention to contraction. The following remarks are easy and well known. We include a short proof for completeness.

Proposition 8.3.4. *The following classes of simplicial complexes are closed under contraction:*

(i) *Paving simplicial complexes;*
(ii) *Matroids.*

Proof. (i) Let $\mathcal{H} = (V, H)$ be paving and let $Q \in H \setminus \{V\}$. Assume that $X \in H/Q$ and $Y \subseteq V \setminus Q$ satisfies $|Y| = |X| - 1$. Then $X \cup Q \in H$ and $|Y \cup Q| = |X \cup Q| - 1$. Since $\mathcal{H}$ is paving, it follows that $X \cup Q \in H$ and so $Y \in H/Q$. Therefore $\mathrm{lk}(Q)$ is paving.

(ii) Let $\mathcal{H} = (V, H)$ be a matroid and let $Q \in H \setminus \{V\}$. Assume that $I, J \in H/Q$ satisfy $|J| = |I| - 1$. Then $I \cup Q, J \cup Q \in H$ and $|I \cup Q| = |J \cup Q| - 1$. Since $\mathcal{H}$ is a matroid, it follows that $J \cup Q \cup \{i\} \in H$ for some $i \in (I \cup Q) \setminus (J \cup Q) = I \setminus J$. Thus $J \cup \{i\} \in H/Q$ and so $\mathrm{lk}(Q)$ is a matroid. $\square$

The remaining classes of simplicial complexes featuring Proposition 8.3.1 behave differently:

Example 8.3.5. The following classes of simplicial complexes are not closed under contraction:

(i) Boolean representable simplicial complexes;
(ii) Graphic boolean simplicial complexes;
(iii) Simplicial complexes satisfying (PR).

Indeed, consider the tetrahedron complex $T_2 = (V, H)$ defined by $V = \{1, \ldots, 4\}$ and $H = P_{\leq 2}(V) \cup \{123, 124\}$. We saw in Example 5.2.11 that T_2 is boolean representable, in fact the computation of the flats there shows that T_2 is graphic boolean. Now $\mathrm{lk}(4) = (\{1, 2, 3\}, \{\emptyset, 1, 2, 3, 12\})$ fails (PR) for 12 and 3. In view of Proposition 5.1.2, T_2 serves as a counterexample for all the three classes above.

The following result, proved by Izhakian and Rhodes, shows that contractions and (PR) together can characterize matroids:

Proposition 8.3.6 ([28, Proposition 2.26(iii)]). *Let $\mathcal{H}$ be a simplicial complex. If all contractions of $\mathcal{H}$ satisfy (PR), then $\mathcal{H}$ is a matroid.*

Proof. Suppose that $\mathcal{H} = (V, H)$ is not a matroid. Then there exist $I, J \in H$ such that $|I| = |J| + 1$ and $J \cup \{i\} \notin H$ for every $i \in I \setminus J$. Let $X = I \cap J$ and assume that $|X|$ is maximal for all possible choices of I, J.

Let $lk(X) = (V \setminus X, H')$. Write $I' = I \setminus X$, $J' = J \setminus X \in H'$. Since $J' = \emptyset$ implies $J \subset I$, which is impossible, we can take $j_0' \in J'$.

Suppose that $lk(X)$ satisfies (PR). Then there exists some $i_0' \in I'$ such that $I'' = (I' \setminus \{i_0'\}) \cup \{j_0'\} \in H'$. Hence $I'' \cup X \in H$. Clearly, $J \cup \{i\} \notin H$ for every $i \in (I'' \cup X) \setminus J \subseteq I \setminus J$. Since $(I'' \cup X) \cap J = X \cup \{j_0'\} \supset X$, this contradicts the maximality of $|X|$. Thus $lk(X)$ fails (PR) and we are done. $\square$

We can also prove the following proposition:

Proposition 8.3.7. *Every simplicial complex is a contraction of some boolean representable simplicial complex.*

Proof. Let $\mathcal{H} = (V, H)$ be a simplicial complex and take a new symbol $z \notin V$. Let $\mathcal{H}' = (V', H')$ be the simplicial complex defined by $V' = V \cup \{z\}$ and

$$H' = \{I \cup \{x\} \mid I \in H, \, x \in V'\} \cup H.$$

It is easy to see that $H \subseteq Fl\ \mathcal{H}'$. Let $X \in H'$ be nonempty. Then there exists some $x \in X$ such that $X \setminus \{x\} \in H$. Since $x \notin X \setminus \{x\}$, which is closed in $\mathcal{H}'$, it follows from Corollary 5.2.7 that $\mathcal{H}'$ is boolean representable.

It is easy to see that $\mathcal{H}$ is a contraction of $\mathcal{H}'$ by verifying that $\mathcal{H} = lk(z)$. $\square$

Unlike the matroid case, negative results appear also when we consider the notion of dual. Given a simplicial complex $\mathcal{H} = (V, H)$, we define the *dual* simplicial complex $\mathcal{H}^* = (V, H^*)$ through the equivalence

$$X \in fct\ \mathcal{H}^* \Leftrightarrow V \setminus X \in fct\ \mathcal{H} \quad (X \subseteq V).$$

It is well known that the dual of a matroid is a matroid [54, Section 5.2]. The next example shows that this property fails for boolean representable simplicial complexes:

Example 8.3.8. The tetrahedron complex T_2 is boolean representable but its dual is not.

Indeed, $T_2 = (1234, P_{\leq 2}(1234) \cup \{123, 124\})$ is boolean representable by Example 5.2.11(iii). We have $fct\ T_2 = \{123, 124, 34\}$, hence $fct\ T_2^* = \{4, 3, 12\}$ and so

$$T_2^* = (1234, P_{\leq 1}(1234) \cup \{12\}). \tag{8.5}$$

Since T_2^* fails (PR) for 12 and 3, it follows from Proposition 5.1.2 that T_2^* is not boolean representable.

In the boolean representable setting, unlike in the matroid case, contraction is not the dual operation of restriction.

Example 8.3.9. Let $V = \{1, \ldots, 4\}$ and $T_2 = (V, P_{\leq 2}(V) \cup \{123, 124\})$. Then the restriction of T_2^* to $V \setminus 13$ is not the dual of the contraction of T_2 by 13.

Indeed, by (8.5), the restriction of T_2^* to $V \setminus 13 = 24$ is $(24, \{\emptyset, 2, 4\})$. On the other hand, the contraction of T_2 by 13 is $(24, \{\emptyset, 2\})$ and has dual $(24, \{\emptyset, 4\})$.

In matroid theory, a minor of $\mathcal{H}$ is any matroid obtained from $\mathcal{H}$ by a sequence of restrictions and contractions. A consequence of Example 8.3.5 is that the theory of forbidden minors for matroids (see [54, Chapter 7]) cannot be generalized to the boolean representable case. In the next section, we show that we can somehow get away with restriction only.

8.4 Prevarieties of Simplicial Complexes

Since *prevarieties* of simplicial complexes are not known in the literature, we dare to define them: classes of simplicial complexes closed under isomorphism and restriction. Hence all the classes in Proposition 8.3.1 constitute prevarieties of simplicial complexes.

Let Σ denote a set of simplicial complexes. We denote by $FR(\Sigma)$ the class of all simplicial complexes having no restriction isomorphic to some element of Σ, so FR stands for *forbidden restriction*.

Given a simplicial complex $\mathcal{H}$ and $k \in \mathbb{N}$, let $Res_k(\mathcal{H})$ denote the set of all restrictions of $\mathcal{H}$ with at most k vertices. We denote by $Res'(\mathcal{H})$ the set of all proper restrictions of $\mathcal{H}$.

Let $\mathcal{SC}$ denote the prevariety of all simplicial complexes. Given a prevariety $\mathcal{V} \subset \mathcal{SC}$ of simplicial complexes, write

$$\tilde{\mathcal{V}} = \{\mathcal{H} \in \mathcal{SC} \setminus \mathcal{V} \mid Res'(\mathcal{H}) \subseteq \mathcal{V}\}.$$

Lemma 8.4.1. *(i) For every set Σ of simplicial complexes, $FR(\Sigma)$ is a prevariety of simplicial complexes.*
(ii) For every prevariety $\mathcal{V}$ of simplicial complexes, $\mathcal{V} = FR(\tilde{\mathcal{V}})$.

Proof. (i) Clearly, $FR(\Sigma)$ is closed under isomorphism. Let $(V, H) \in FR(\Sigma)$ and let $V' \subseteq V$, $H' = H \cap 2^{V'}$. Suppose that (V', H') has a restriction (V'', H'') isomorphic to some element of Σ. Then (V'', H'') is itself a restriction of (V, H), contradicting $(V, H) \in FR(\Sigma)$. Thus $FR(\Sigma)$ is closed under restriction and constitutes therefore a prevariety.
(ii) The inclusion $\mathcal{V} \subseteq FR(\tilde{\mathcal{V}})$ is immediate. Conversely, if $\mathcal{H} \in \mathcal{SC} \setminus \mathcal{V}$, then $\mathcal{H}$ must have some restriction in $\tilde{\mathcal{V}}$, hence $\mathcal{H} \notin FR(\tilde{\mathcal{V}})$. $\square$

If $\mathcal{V} = FR(\Sigma)$, we say that Σ is a *basis* of $\mathcal{V}$. We say that $\mathcal{V}$ is *finitely based* if it admits a finite basis. The *size* of a nonempty basis is

$$\sup\{|V| \mid (V, H) \in \Sigma\}.$$

By convention, the size of the empty basis is 0 (note that $FR(\emptyset)$ is the class of all simplicial complexes). We say that $\mathcal{V}$ has size $k \in \mathbb{N} \cup \{\infty\}$ (and write $siz \mathcal{V} = k$) if

$\mathcal{V}$ admits a basis of size k, but not smaller. Since we do not need to keep isomorphic simplicial complexes in a basis, $\mathcal{V}$ is finitely based if and only if siz $\mathcal{V} < \infty$.

The following example shows that prevarieties need not be finitely based, even if their elements have bounded dimension. In Sect. 4.1.2, we identified (finite undirected) graphs with trim simplicial complexes of dimension ≤ 1.

Example 8.4.2. Let $\mathcal{A}$ denote the class of all finite undirected acyclic graphs. Then $\mathcal{A}$ is a non finitely based prevariety of simplicial complexes.

Indeed, since a subgraph of an acyclic graph is necessarily acyclic, then $\mathcal{A}$ is a prevariety of simplicial complexes. Suppose that $\mathcal{A} = \mathrm{FR}(\Sigma)$ for some finite set Σ of simplicial complexes. Let k be the size of Σ and let $\mathcal{H}$ be a graph consisting of a single cycle of length $k + 1$. Since $\mathcal{H} \notin \mathcal{A}$, then $\mathcal{H}$ must have some restriction $\mathcal{H}'$ isomorphic to some element of Σ. But then $\mathcal{H}'$ has at most k vertices and so $\mathcal{H}' \in \mathcal{A}$, contradicting $\mathcal{A} = \mathrm{FR}(\Sigma)$. Therefore $\mathcal{A}$ is not finitely based.

We complete this section by proving two results that will help us to compute sizes in Sect. 8.5:

Lemma 8.4.3. *Let* Σ, Σ' *be classes of simplicial complexes and let* $\mathcal{V}, \mathcal{V}'$ *be prevarieties of simplicial complexes. Then:*

(i) $\mathrm{FR}(\Sigma) \cap \mathrm{FR}(\Sigma') = \mathrm{FR}(\Sigma \cup \Sigma')$;
(ii) $\mathrm{siz}(\mathcal{V} \cap \mathcal{V}') \leq \max\{\mathrm{siz}\,\mathcal{V}, \mathrm{siz}\,\mathcal{V}'\}$;
(iii) *If* $\mathcal{V}, \mathcal{V}'$ *are finitely based, so is* $\mathcal{V} \cap \mathcal{V}'$.

Proof. (i) is straightforward, (ii) follows from (i) and (iii) from (ii). $\square$

Theorem 8.4.4. *Let* $\mathcal{V} \subset \mathcal{SC}$ *be a prevariety of simplicial complexes. Then*

$$\mathrm{siz}\,\mathcal{V} = \sup\{|V| \mid (V, H) \in \tilde{\mathcal{V}}\}.$$

Proof. By Lemma 8.4.1(ii), $\tilde{\mathcal{V}}$ is a basis of $\mathcal{V}$, hence

$$\mathrm{siz}\,\mathcal{V} \leq \sup\{|V| \mid (V, H) \in \tilde{\mathcal{V}}\}.$$

Suppose that $\mathcal{V} = \mathrm{FR}(\Sigma)$ with siz $\Sigma = $ siz $\mathcal{V}$. Let $\mathcal{H} = (V, H) \in \tilde{\mathcal{V}}$. Then $\mathcal{H} \notin \mathrm{FR}(\Sigma)$ and so $\mathcal{H}$ has some restriction isomorphic to some complex in Σ. Since $\mathrm{Res}'(\mathcal{H}) \subseteq \mathcal{V}$, it follows that $\mathcal{H}$ itself is isomorphic to some complex in Σ and so

$$|V| \leq \sup\{|V'| \mid (V', H') \in \Sigma\} = \mathrm{siz}\,\Sigma = \mathrm{siz}\,\mathcal{V}.$$

Therefore

$$\sup\{|V| \mid (V, H) \in \tilde{\mathcal{V}}\} \leq \mathrm{siz}\,\mathcal{V}$$

and we are done. $\square$

8.5 Finitely Based Prevarieties

We introduce now some notation for prevarieties of simplicial complexes:

- $\mathcal{PV}$: paving simplicial complexes;
- $\mathcal{PR}$: simplicial complexes satisfying (PR);
- $\mathcal{BR}$: boolean representable simplicial complexes;
- $\mathcal{GB}$: graphic boolean simplicial complexes;
- $\mathcal{MT}$: matroids;
- $\mathcal{PM} = \mathcal{MT} \cap \mathcal{PV}$;
- $\mathcal{PB} = \mathcal{PV} \cap \mathcal{BR}$.

Given a prevariety $\mathcal{V}$ of simplicial complexes and $d \in \mathbb{N}$, we define also the prevariety

$$\mathcal{V}_d = \{\mathcal{H} \in \mathcal{V} \mid \dim \mathcal{H} \le d\}.$$

In general, we need to consider dimension restrictions to obtain finitely based prevarieties. But Example 8.4.2 shows that dimension restrictions do not imply finitely based.

Theorem 8.5.1. *Let $d \ge 1$. Then*

- *(i)* $\operatorname{siz} \mathcal{PV}_d = 2d + 1$;
- *(ii)* $\operatorname{siz} \mathcal{PR}_d = d + 2$;
- *(iii)* $\operatorname{siz} \mathcal{MT}_d = \operatorname{siz} \mathcal{PM}_d = d + 2$;
- *(iv)* $\operatorname{siz} \mathcal{GB} = 6$.

Proof. (i) Let $\mathcal{H} = (V, H) \in \mathcal{P}\tilde{\mathcal{V}}_d$. If $\dim \mathcal{H} > d$, then $U_{d+2,d+2} \notin \mathcal{PV}_d$ is a restriction of $\mathcal{H}$ and so $|V| = d + 2$ by minimality. Hence we may assume that $\dim \mathcal{H} = r \le d$. It follows that there exist $I \in H \cap P_{r+1}(V)$ and $J \in P_r(V) \setminus H$. Thus $(I \cup J, H \cap 2^{I \cup J}) \notin \mathcal{PV}_d$ and by minimality we get $I \cup J = V$. Thus $|V| \le |I| + |J| = 2r + 1 \le 2d + 1$ and so $\operatorname{siz} \mathcal{PV}_d \le 2d + 1$ by Theorem 8.4.4.

Equality comes from presenting some $(V, H) \in \mathcal{P}\tilde{\mathcal{V}}_d$ with $|V| = 2d + 1$. We take $V = \{0, \ldots, 2d\}$ and

$$H = (P_{\le d}(V) \setminus \{\{d + 1, d + 2, \ldots, 2d\}\}) \cup \{01 \ldots d\}.$$

Then $(V, H) \notin \mathcal{PV}_d$. Let $V' \subseteq V$ with $|V'| = 2d$ and write $H' = H \cap 2^{V'}$. If $01 \ldots d \subseteq V'$, then

$$H' = P_{\le d}(V') \cup \{01 \ldots d\}.$$

If $01 \ldots d \not\subseteq V'$, then

$$H' = P_{\le d}(V') \setminus \{(d + 1)(d + 2) \ldots (2d)\}$$

and so in any case we get $(V', H') \in \mathcal{PV}_d$. Therefore $(V, H) \in \mathcal{P}\tilde{\mathcal{V}}_d$ as desired.

(ii) Let $\mathcal{H} = (V, H) \in \mathcal{P\tilde{R}}_d$. If dim $\mathcal{H} > d$, then $U_{d+2,d+2} \notin \mathcal{PR}_d$ is a restriction of $\mathcal{H}$ and so $|V| = d + 2$ by minimality. Hence we may assume that dim $\mathcal{H} = r \le d$. Since $\mathcal{H} \notin \mathcal{PR}_d$, there exist $I, \{p\} \in H$ such that $(I \setminus \{i\}) \cup \{p\} \notin H$ for every $i \in I$. It follows that the restriction induced by $I \cup \{p\}$ is not in $\mathcal{PR}$ either, hence siz $\mathcal{PR}_d \le d + 2$.

Equality follows from noting that $U_{d+2,d+2} \in \mathcal{P\tilde{R}}_d$.

(iii) Let $\mathcal{H} = (V, H) \in \mathcal{M\tilde{T}}_d$. If dim $\mathcal{H} > d$, then $U_{d+2,d+2} \notin \mathcal{MT}_d$ is a restriction of $\mathcal{H}$ and so $|V| = d + 2$ by minimality. Hence we may assume that dim $\mathcal{H} = r \le d$.

Let $I, J \in H$ fail (EP). By minimality of $|V|$, every proper restriction of $\mathcal{H}$ is a matroid, whence $I \cup J = V$. Let $a \in J \setminus I$ and let (V', H') be the restriction of (V, H) determined by $V' = V \setminus \{a\}$. Applying (EP') twice in succession to $I, J' = J \setminus \{a\} \in H'$, it follows that there exist distinct $i, i' \in I \setminus J$ such that $I' = J' \cup \{i, i'\} \in H'$. Now (EP) fails also for I' and J, and by minimality of $|V|$ we must have $I' \cup J = V$. Since $|I' \setminus J| = 2$, we get $|V| = |J| + 2 = r + 2 \le d + 2$. Therefore siz $\mathcal{MT}_d \le d + 2$ by Theorem 8.4.4.

Equality now follows from $U_{d+2,d+2} \in \mathcal{M\tilde{T}}_d$. With the same proof, we get also the equality siz $\mathcal{PM}_d = d + 2$.

(iv) In view of Proposition 6.2.1 and Theorem 8.4.4, we have siz $\mathcal{GB} \le 6$.

Now let $V = \{1, \ldots, 6\}$ and $H = P_{\le 3}(V) \setminus \{124, 135, 236\}$. Note that H is obtained by excluding from $P_{\le 3}(V)$ the lines of the PEG

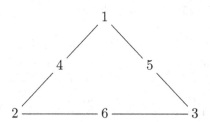

By Proposition 6.2.1, (V, H) is not graphic boolean. Let $V' \subseteq V$ with $|V'| = 5$ and write $H' = H \cap 2^{V'}$.

Then $P_{\le 3}(V') \setminus H'$ has at most two elements. Since the elements abx, ayc and zbc in the statement of Proposition 6.2.1 must be all distinct, it follows that $(V', H') \in \mathcal{GB}$. Hence $(V, H) \in \mathcal{\tilde{G}B}$ and so siz $\mathcal{GB} = 6$ by Theorem 8.4.4. $\square$

Boolean representability is a tougher challenge. We start with the paving case before facing the harder general case.

Theorem 8.5.2. *(i)* siz $\mathcal{PB}_1 = 3$.
(ii) For every $d \ge 2$, siz $\mathcal{PB}_d = (d + 1)(d + 2)$.

Proof. (i) Let $\mathcal{H} = (V, H) \in \mathcal{P}\tilde{\mathcal{B}}_1$. If dim $\mathcal{H} > 1$, we get $|V| = 3$ as in the proof of Theorem 8.5.1(i). On the other hand, Corollary 5.3.2 excludes dim $\mathcal{H} \leq 0$, hence we may assume that dim $\mathcal{H} = 1$. If $\mathcal{H}$ is not paving, we may proceed as in the proof of Theorem 8.5.1(i) to get $|V| = 3$, hence we assume that $\mathcal{H} \notin \mathcal{BR}$. By Proposition 5.3.1, there exist $a, b, c \in V$ such that $ab, bc \notin H$ but $ac \in H$. It follows also from Proposition 5.3.1 that the restriction of $\mathcal{H}$ induced by $V' = abc$ is not in $\mathcal{BR}$ either. Thus $V = V'$ and so siz $\mathcal{P}\mathcal{B}_1 = 3$.

(ii) Let $\mathcal{H} = (V, H) \in \mathcal{P}\tilde{\mathcal{B}}_d$. If dim $\mathcal{H} > d$, we get $|V| = d + 2 \leq (d + 1)(d + 2)$ as in the proof of Theorem 8.5.1(i). On the other hand, if $\mathcal{H}$ is not paving we get $|V| \leq 2d + 1 \leq (d + 1)(d + 2)$ by Theorem 8.5.1(i). Hence we may assume that $\mathcal{H} \in \mathcal{PV}_d$ and so $\mathcal{H} \notin \mathcal{BR}$. Since the function $(d + 1)(d + 2)$ is increasing for $d \geq 2$, we may assume that dim $\mathcal{H} = d$. By Proposition 6.1.2, there exists some $A \in H \cap P_{d+1}(V)$ such that $\overline{A \setminus \{a\}} = V$ for every $a \in A$. Assume that $A = a_0 \ldots a_d$ and write $A_i = A \setminus \{a_i\}$. For each $i = 0, \ldots, d$, let $S_i \subseteq V$ be maximal for the properties

- $A_i \subseteq S_i$;
- $P_{d+1}(S_i) \cap H = \emptyset$.

Note that A_i would satisfy both conditions above, hence there exists some maximal S_i satisfying them. We claim that, for $i = 1, \ldots, m$,

$$\exists x_{i0} \in V \setminus S_i \ \exists x_{i1}, \ldots, x_{i,2d} \in S_i :$$
$$(x_{i0}x_{i1} \ldots x_{id} \in H \quad \text{and} \quad x_{i0}x_{i,d+1} \ldots x_{i,2d} \notin H). \tag{8.6}$$

Indeed, since dim $\mathcal{H} = d$ we have $S_i \subset V = \overline{A_i}$. Hence S_i is not closed. Since $H \cap P_{d+1}(S_i) = \emptyset$ and $P_{\leq d}(V) \subseteq H$, it follows that there exist distinct $x_{i,d+1}, \ldots, x_{i,2d} \in S_i$ and $x_{i0} \in V \setminus S_i$ such that $x_{i0}x_{i,d+1} \ldots x_{i,2d} \notin H$. On the other hand, since $x_{i0} \notin S_i$, it follows from maximality of S_i that $x_{i0}x_{i1} \ldots x_{id} \in H$ for some distinct $x_{i1}, \ldots, x_{id} \in S_i$. Thus (8.6) holds.

Note that, in view of our notational conventions, we are assuming $x_{i1}, \ldots, x_{id}$ to be distinct and $x_{i,d+1}, \ldots, x_{i,2d}$ to be distinct as well, but these two sets may intersect each other. Therefore, we may assume that

$$x_{i1} \ldots x_{id} = A_i \quad \text{or} \quad x_{i,d+1} \ldots x_{i,2d} = A_i. \tag{8.7}$$

Let (V', H') be the restriction of $\mathcal{H}$ induced by the subset

$$V' = \{x_{ij} \mid i = 0, \ldots, d; \ j = 0, \ldots, 2d\}.$$

In view of (8.7), we have $A \in H'$. Let $\text{Cl}'X$ denote the closure of $X \subseteq V'$ in (V', H') and let $i \in \{0, \ldots, d\}$. Given $s \in S_i \setminus A_i$, we have $A_i \cup \{s\} \notin H'$ by definition of S_i, hence $A_i \in H'$ yields $S_i \subseteq \text{Cl}'A_i$. Now $x_{i,d+1} \ldots x_{i,2d} \in H' \cap 2^{S_i}$ and $x_{i0}x_{i,d+1} \ldots x_{i,2d} \notin H'$ together yield $x_{i0} \in \text{Cl}'A_i$ and so $\text{Cl}'A_i$ contains the $(d + 1)$-set $x_{i0}x_{i1} \ldots x_{i,d} \in H'$. By Proposition 4.2.3, we get $\text{Cl}'A_i = V'$ for $i = 1, \ldots, m$. By Proposition 6.1.2, (V', H') is not boolean representable. By minimality of $|V|$, we get $V' = V$.

Now (8.7) implies that each a_k occurs d times among the x_{ij}. Hence $|V| = |V'| \leq (d + 1)(2d + 1) - (d + 1)(d - 1) = (d + 1)(d + 2)$ and so $\text{siz}\,\mathcal{PB}_d \leq (d + 1)(d + 2)$ by Theorem 8.4.4.

To prove equality, we build some $(V, H) \in \mathcal{P}\tilde{\mathcal{B}}_d$ with $|V| = (d + 1)(d + 2)$.

Let $A = \{a_0, \ldots, a_d\}$ and $B_i = \{b_{i0}, \ldots, b_{id}\}$ for $i = 0, \ldots, d$. Write also $A_i = A \setminus \{a_i\}$ and

$$C_i = P_{d+1}(A_i \cup (B_i \setminus \{b_{i0}\})) \cup \{B_i\}.$$

We define

$$V = A \cup \bigcup_{i=0}^{d} B_i, \quad H = P_{\leq d+1}(V) \setminus \bigcup_{i=0}^{d} C_i.$$

Clearly, $|V| = (d + 1)(d + 2)$. We show that (V, H) is not boolean representable.

We have $A \in H$. Let $i \in \{0, \ldots, d\}$. Since $H \cap P_{d+1}(A_i \cup (B_i \setminus \{b_{i0}\})) = \emptyset$, it is easy to see that $B_i \setminus \{b_{i0}\} \subseteq \overline{A_i}$ and so $B_i \setminus \{b_{i0}\} \in H \cap 2^{\overline{A_i}}$ together with $B_i \notin H$ yields also $b_{i0} \in \overline{A_i}$. Thus $A_i \cup \{b_{i0}\} \in H \cap 2^{\overline{A_i}}$ and so $\overline{A_i} = V$ by Proposition 4.2.3. By Proposition 6.1.2, (V, H) is not boolean representable and so $(V, H) \notin \mathcal{PB}_d$.

Consider now $v \in V$ and let (V', H') be the restriction of (V, H) induced by $V' = V \setminus \{v\}$. Clearly, $(V', H') \in \mathcal{PV}_d$, so all we need is to prove that $(V', H') \in \mathcal{BR}$. Let $\text{Cl}'X$ denote the closure of $X \subseteq V'$ in (V', H'). By Proposition 6.1.2, it suffices to show that, for every $X \in H' \cap P_{d+1}(V')$, there exists some $x \in X$ such that $x \notin \text{Cl}'(X \setminus \{x\})$. We consider three cases.

Assume first that $X \in P_{d+1}(A_i \cup B_i)$. Then $b_{i0} \in X$ and $a_j \in X$ for some $j \neq i$. Taking $x \in X \setminus \{b_{i0}, a_j\}$, it follows easily that $X \setminus \{x\}$ is closed in (V', H') and so $x \notin \text{Cl}'(X \setminus \{x\})$.

Assume next that X contains simultaneously elements b_{ij} and $y \notin A_i \cup B_i$. Taking $x \in X \setminus \{b_{ij}, y\}$, it is easy to check that $X \setminus \{x\}$ is closed in (V', H') and so $x \notin \text{Cl}'(X \setminus \{x\})$.

Therefore we may assume that $X = A$. It follows that $v \in B_i$ for some i. Let $x = a_i$. We claim that $x \notin \text{Cl}'(X \setminus \{x\}) = \text{Cl}'A_i$. Once again, we get $B_i \setminus \{b_{i0}, v\} \subseteq \text{Cl}'A_i$. Since $B_i \not\subseteq V'$, it is straightforward to check that $A_i \cup (B_i \setminus \{b_{i0}, v\})$ is closed in (V', H'), hence $x \notin \text{Cl}'(X \setminus \{x\})$ in all three cases and we are done. $\square$

If we drop the paving restriction, things get more complicated, but we still have a finitely based prevariety. We need the following lemma:

Lemma 8.5.3. *Let* $\mathcal{H} = (V, H)$ *be a simplicial complex of dimension* $d \geq 1$ *and let* $X \subseteq V$, $a \in \overline{X}$. *Then there exists a restriction* (V', H') *of* $\mathcal{H}$ *such that* $|V'| \leq (d + 1)d^{2d} + 1$ *and* a *belongs to the closure of* $X \cap V'$ *in* (V', H').

Proof. Given $Y \subseteq Z \subseteq E$, let $\mathrm{Cl}_Z Y$ denote the closure of Y in the restriction $(Z, H \cap 2^Z)$. We define a finite sequence $X_0, X_1, \ldots, X_n$ of disjoint subsets of V as follows. Let $X_0 = X$. Assume that $X_0, X_1, \ldots, X_{k-1}$ are defined. If $X_{k-1} = \emptyset$ or $X'_{k-1} = X_0 \cup \ldots \cup X_{k-1}$ contains a facet of $\mathcal{H}$, the sequence stops at X_{k-1}. Otherwise, let

$$X_k = \{b \in V \setminus X'_{k-1} \mid Y \cup \{b\} \notin H \text{ for some } Y \in H \cap 2^{X'_{k-1}}\}.$$

We prove that $n \leq 2d$.

We may assume that $X \neq \emptyset$. Suppose first that $\{x\} \notin H$ for every $x \in X$. Then $X_1 = \{x \in V \setminus X \mid \{x\} \notin H\}$. If $X_1 \neq \emptyset$, then $X_2 = \emptyset$, yielding $n \leq 2 \leq 2d$. Hence we may assume that there exists some $a_0 \in X \cap H$.

Suppose that $n > 2d$. Let $a_i \in X_{2i}$ for $i = 1, \ldots, d$ and write $A_i = a_0 a_1 \ldots a_i \in X'_{2i}$. Since $A_0 \in H$, we can define k to be the maximum value of A_i such that $A_i \in H$. Suppose that $i < d$. Then $A_{i+1} = A_i \cup \{a_{i+1}\} \notin H$ and so $a_{i+1} \in X_{2i+1}$, contradicting $a_{i+1} \in X_{2(i+1)}$. It follows that $A_d \in H$ and so X'_{2d} contains a facet, contradicting $n > 2d$. Thus $n \leq 2d$.

Next we show that, for all $i \leq n$ and $p \in X_i$,

$$p \in \mathrm{Cl}_{V_p}(X \cap V_p) \text{ for some } V_p \subseteq V \text{ such that } |V_p| \leq d^i. \tag{8.8}$$

We use induction on i. If $i = 0$, we take $V_p = \{p\}$ and the claim holds. Assume now that $i > 0$ and the claim holds for smaller values of i. Since $p \in X_i$, there exists some $Y \in H \cap 2^{X'_{i-1}}$ such that $Y \cup \{p\} \notin H$. Since $|Y| \leq d$ (otherwise Y would be a facet and X_i would not be defined), and using the induction hypothesis, we find for each $y \in Y$ some $V_y \subseteq V$ such that $y \in \mathrm{Cl}_{V_y}(X \cap V_y)$ and $|V_y| \leq d^{i-1}$. Writing $V_p = \cup_{y \in Y} V_y$, we claim that $p \in \mathrm{Cl}_{V_p}(X \cap V_p)$. Indeed, it follows from Proposition 8.3.3(ii) that $Y \subseteq \mathrm{Cl}_{V_p}(X \cap V_p)$. Since $Y \in H$ and $Y \cup \{p\} \notin H$, we get $p \in \mathrm{Cl}_{V_p}(X \cap V_p)$. Since $|V_p| \leq \sum_{y \in Y} |V_y| \leq d \cdot d^{i-1} = d^i$, then (8.8) holds.

We consider now two cases. Suppose first that $a \in X'_n$. Since $n \leq 2d$, we may apply (8.8) directly to prove the claim of the lemma. Hence we may assume that $a \notin X'_n$. Note that the sequence $(X_i)_i$ terminates due to one of two reasons. Either $X_n = \emptyset$ or X'_n contains a facet of $\mathcal{H}$. Suppose that $X_n = \emptyset$. It follows easily that X'_n is closed, contradicting $a \in \overline{X}$ in view of $a \notin X'_n$. Thus X'_n must contain a facet Y of $\mathcal{H}$. Since $|Y| \leq d + 1$, we can apply (8.8) to each of the elements of Y and take $V' = \{a\} \cup (\cup_{y \in Y} V_y)$ as in the proof of (8.8). Therefore $a \in V' = \mathrm{Cl}_{V'}(X \cap V')$ and $|V'| \leq (d+1)d^n + 1 \leq (d+1)d^{2d} + 1$ as required. $\square$

Theorem 8.5.4. *(i)* $\mathrm{siz}\, \mathcal{BR}_1 = 3$.
(ii) $\mathrm{siz}\, \mathcal{BR}_2 = 12$.
(iii) For every $d \geq 3$, $\mathrm{siz}\, \mathcal{BR}_d \leq (d+1)^2 d^{2d} + d + 1$.

Proof. (i) This follows immediately from the proof of Theorem 8.5.2(i).

(ii) Let $\mathcal{H} = (V, H) \in \tilde{\mathcal{BR}}_2$. Similarly to the proof of Theorem 8.5.1(i), we may assume that $\dim \mathcal{H} = 2$ and so $\mathcal{H} \notin \mathcal{BR}$. By Corollary 5.2.7, there exists some nonempty $A \in H$ such that $a \in \overline{A \setminus \{a\}}$ for every $a \in A$.

If $|A| = 3$, then we adapt the proof of Theorem 8.5.2(ii), defining the subsets S_i exactly the same way. We claim that, for $i = 1, 2, 3$,

$$\exists x_{i0} \in V \setminus S_i \quad \exists x_{i1}, \ldots, x_{i,4} \in S_i : \tag{8.9}$$
$$(x_{i0} x_{i1} x_{i2} \in H \quad \text{and} \quad x_{i0} x_{i3} x_{i4} \notin H).$$

Indeed, since $a_i \notin S_i$ we have $S_i \subset \overline{A_i}$. Hence S_i is not closed. Hence there exists some $Y_i \subseteq H \cap 2^{S_i}$ and some $x_{i0} \in V \setminus S_i$ such that $Y_i \cup \{x_{i0}\} \notin H$. Since $P_3(S_i) \cap H = \emptyset$, we have $|Y_i| \leq 2$. Completing the elements of Y_i with others if necessary, it follows that there exist distinct $x_{i3}, x_{i4} \in S_i$ such that $x_{i0} x_{i3} x_{i4} \notin H$. On the other hand, since $x_{i0} \notin S_i$, it follows from the maximality of S_i that $x_{i0} x_{i1} x_{i2} \in H$ for some distinct $x_{i1}, x_{i2} \in S_i$. Thus (8.9) holds.

Now we mimic the proof of Theorem 8.5.2(ii) to get $|V| \leq 12$ in this case.

It remains to be considered the case $|A| = 2$, say $A = ab$. Suppose that all the connected components of the graph $\Gamma\,\mathcal{H}$ are cliques. Let C_a denote the vertices in the connected component of a. Similarly to the proof of Proposition 5.3.1, we have

$$\overline{a} = (V \setminus H) \cup C_a,$$

hence $b \notin \overline{a}$, a contradiction. Hence there exist distinct edges $x \,\text{---}\, y \,\text{---}\, z$ in $\Gamma\,\mathcal{H}$ such that x is not adjacent to z, i.e. $xz \in H$. Taking $V' = \{x, y, z\}$, it follows from Proposition 5.3.1 that the restriction induced by V' is not boolean representable and so $|V| = 3$ by minimality.

From both cases we deduce siz $\mathcal{BR}_2 \leq 12$. For equality, we may of course use the same simplicial complex as in the proof of Theorem 8.5.2(ii).

(iii) Let $\mathcal{H} = (V, H)$ be a simplicial complex not in $\mathcal{BR}_d$. Without loss of generality, we may assume that dim $\mathcal{H} = d$: if dim $\mathcal{H} > d$, then $\mathcal{H}$ has a restriction $U_{d+2,d+2}$, and the case dim $\mathcal{H} < d$ is a consequence of the case dim $\mathcal{H} = d$ since our bound increases with d.

By Corollary 5.2.7, there exists some $A \in H$ such that $a \in \overline{A \setminus \{a\}}$ for every $a \in A$. By Lemma 8.5.3, for every $a \in A$ there exists a restriction (V_a, H_a) of $\mathcal{H}$ such that $|V_a| \leq (d+1)d^{2d} + 1$ and a belongs to the closure of $(A \setminus \{a\}) \cap V_a$ in (V_a, H_a). Now we take $V' = \cup_{a \in A} V_a$. By Proposition 8.3.3(ii), a belongs to the closure of $A \setminus \{a\}$ in $(V', H \cap 2^{V'})$. Since

$$|V'| \leq (d+1)|V_a| \leq (d+1)^2 d^{2d} + d + 1,$$

we get siz $\mathcal{BR}_d \leq (d+1)^2 d^{2d} + d + 1. \quad \square$

Chapter 9
Open Questions

As a general objective we would like to raise the results in this monograph from dimension 2 to dimension 3 and further.

Below we will list more specific questions on the representation/combinatorial/geometric/topological theories of $\mathcal{BR}$ and on the theory of finite posets.

Many of these questions remain new and important when restricted to matroids.

9.1 Representation Theory of $\mathcal{BR}$ and Matroids

Question 9.1.1. *Given* $\mathcal{H} \in \mathcal{BR}$, *is* mindeg $\mathcal{H}$ *always achieved by an sji (a minimal) lattice representation of* $\mathcal{H}$?

Question 9.1.2. *Is there additional structure on the set of boolean representations of a given* $\mathcal{H} \in \mathcal{BR}$ *beyond the structure with join "stacking of boolean matrices" detailed in this monograph? For example, is there a tensor product of semilattices structure?*

Question 9.1.3. *Calculate the minimal/sji representations and* mindeg *for the following matroids:*

(i) *All projective planes (we did the Fano plane in Sect. 5.7.2);*

(ii) *Dowling geometries of arbitrary rank for every finite group (see [17]);*

(iii) *The uniform matroids* $U_{m,n}$ *for* $3 \leq m < n$ *(we did* $U_{3,n}$ *in Sect. 5.7.3);*

(iv) $(V, H(\mathcal{B}))$ *for a Steiner system* $(V, \mathcal{B}) \in S(r-1, r, n)$ *(see Theorem 5.7.18(iii) for* $r = 4$ *and* $n = 8$*);*

(v) *Every example of matroid with at most 12 points at the back of [39].*

© Springer International Publishing Switzerland 2015
J. Rhodes, P.V. Silva, *Boolean Representations of Simplicial Complexes and Matroids*, Springer Monographs in Mathematics,
DOI 10.1007/978-3-319-15114-4_9

Question 9.1.4. *Using [28, Theorem 5.4] and [21], apply the boolean representation theory of matroids to get conditions on when a matroid has a matrix representation over a given finite field F.*

Following the proof of [28, Theorem 5.4], we note that a boolean representation of a matroid $\mathcal{H} = (V, H)$ can be constructed from a field representation of $\mathcal{H}$ by stacking matrices. We describe briefly this procedure.

Let $d = \dim \mathcal{H}$ and $n = |V|$. Assume that $M = (a_{ij})$ is an $m \times n$ matrix over a field $\mathbb{F}$ representing $\mathcal{H}$, so that H is precisely the set of all subsets of independent column vectors of M (over $\mathbb{F}$). Since $d = \dim \mathcal{H}$, the matrix M has rank $d + 1$. By performing standard row operations on M such as adding to a row a multiple of another row, we may replace M by a matrix having precisely $d + 1$ nonzero rows, and producing the same subsets of independent column vectors. Hence we may assume that $m = d + 1$.

Let $M' = (b_{ij})$ be an $m' \times n$ boolean matrix satisfying the following condition: $(c_1, \ldots, c_n)$ is a nonzero row vector of M' if and only if there exists a linear combination of row vectors of M having zero entries at the same positions than $(c_1, \ldots, c_n)$. Let $X \subseteq \{1, \ldots, n\}$. We claim that

$$\begin{aligned} \{M[\_, j] \mid j \in X\} \text{ is independent over } \mathbb{F} \text{ if and only if} \\ \{M'[\_, j] \mid j \in X\} \text{ is independent.} \end{aligned} \tag{9.1}$$

Assume that $\{M[\_, j] \mid j \in X\}$ is independent over $\mathbb{F}$. Then there exists some subset Y of $|X|$ rows such that $M[Y, X]$ has nonzero determinant. By adding to a row a multiple of another row, we may transform $M[Y, X]$ into a matrix N congruent to a lower triangular matrix. Each row of N is a (nonzero) linear combination of rows in $M[Y, X]$. Computing these same linear combinations for the full matrix M provides a set Y' of rows in M' such that $M'[Y', X]$ is congruent to a lower unitriangular matrix. Thus $\{M'[\_, j] \mid j \in X\}$ is independent.

Conversely, assume that $\{M'[\_, j] \mid j \in X\}$ is independent. Then there exists a subset Y of $|X|$ rows of M' such that $M'[Y', X]$ is congruent to a lower unitriangular matrix. Each row $M'[i, \_]$ ($i \in Y'$) arises from a linear combination C_i of row vectors of M. Let M'' be the matrix over $\mathbb{F}$ obtained by adding to M the row vectors C_i ($i \in Y'$). Since $M''[Y', X]$ has clearly nonzero determinant, then $\{M''[\_, j] \mid j \in X\}$ is independent over $\mathbb{F}$. Since adding linear combinations or rows does not alter independence of column vectors, it follows that $\{M[\_, j] \mid j \in X\}$ is independent over $\mathbb{F}$ and so (9.1) holds.

Thus M' is a boolean matrix representation of $\mathcal{H}$. Now, being a matroid, $\mathcal{H}$ is pure, so we only need to put enough rows into M' to make the facets of $\mathcal{H}$ independent. If $\mathcal{H}$ has ℓ facets, it follows easily from the above algorithm that we need at most $m\ell$ rows in M'. Therefore

$$\text{mindeg } \mathcal{H} \leq (\dim \mathcal{H} + 1)|\text{fct } \mathcal{H}|.$$

Since not every matroid is field representable, we know that there is no general method for reverting this process. But can it be done for particular subclasses of matroids? We intend to develop these connections in the future.

Question 9.1.5. *Look at [9, 36, 41] and apply the boolean representation theory of matroids to the theory of Coxeter matroids and Bruhat orders.*

9.2 Combinatorial Theory of $\mathcal{BR}$ and Matroids

Question 9.2.1. *Provide better or sharp bounds for* siz $\mathcal{BR}_d$ *(see Theorem 8.5.4(iii)).*

Question 9.2.2. *Extend the analysis of* mindeg *for* BPav(2) *(Theorem 6.5.1):*

(i) To the case where $\Gamma Fl\, \mathcal{H}$ is connected;
(ii) To BPav(3) *and higher dimensions.*

Question 9.2.3. *Discover the structure of all simplicial complexes of dimension d which are not in $\mathcal{BR}$ but have all proper restrictions in $\mathcal{BR}$.*

Question 9.2.4. *Is there some generalization of the greedy algorithm characterization of matroids which applies to $\mathcal{BR}$?*

Question 9.2.5. *We say that a simplicial complex $\mathcal{H}$ is a boolean module of type $B^{(n)}$ if it admits an $n \times (2^n - 1)$ boolean matrix representation where all columns are distinct and nonzero.*

(i) Calculate the independent sets of $\mathcal{H}$.
(ii) Calculate the flats of $\mathcal{H}$.
(iii) Relate to known combinatorial objects in the literature.

For $n = 3$ the independent sets are all the sets with at most two elements plus 25 3-sets, see Example 6.3.5 for the list of the dependent 3-sets.

9.3 Geometric Theory of $\mathcal{BR}$ and Matroids

Question 9.3.1. *Extend the geometric analysis of* BPav(2) *in Sects. 6.3 and 6.4 to* BPav(3) *and higher dimensions, especially Lemma 6.3.3 and Theorem 6.3.4, for geometrically computing independent sets and flats, respectively.*

Question 9.3.2. *Generalize ΓM to* BPav(3) *and then generalize the results of Sects. 6.3 and 6.4 which use ΓM.*

9.4   Topological Theory of $\mathcal{BR}$ and Matroids

Question 9.4.1. *Let* $\mathcal{H} = (V, H) \in$ BPav(2) *with at most one nontrivial connected component in* ΓFl $\mathcal{H}$. *Which total orders on* V *produce a shelling of* $\mathcal{H}$ *through the alphabetic order?*

Question 9.4.2. *Extend the shellability results for* BPav(2) *to* BPav(3) *and beyond, defining the appropriate connectivity conditions which generalize the conditions using* ΓFl $\mathcal{H}$ *in Chap. 7.*

Question 9.4.3. *Do the concepts of shellable and sequentially Cohen-Macaulay coincide for all (paving) boolean representable simplicial complexes? If not, do they coincide for some nice subclass?*

Question 9.4.4. *Have boolean representable simplicial complexes shown up in the topological literature before? If so, where?*

One of our current lines of research indicates that boolean representability can provide important information on the homotopy groups of the complex, at least on the fundamental group. We recall that any finitely presented group can occur as the fundamental group of some simplicial complex of dimension 2.

9.5   Applications of the Theory to Finite Posets

Question 9.5.1. *Sections 3.2 and A.4 develop the theory of boolean representations of finite posets through the Dedekind-MacNeille completion. Develop this theory of c-independence for finite posets along the lines of this monograph. Solve Question 9.1.5 in the poset setting.*

Question 9.5.2. *Given any universal algebra and a definition of substructure closed under all intersections (e.g. groups and subgroups, or semigroups and subsemigroups or ideals), compute the c-independent sets with respect to the lattice of the substructures, discussing dimension (see [11]), shellability, etc.*

Appendix A

We collect in this appendix complementary material of two types:

- Classical results which contribute to making this monograph self-contained;
- Related subjects which may be of interest for future research.

A.1 Supertropical Semirings

As an alternative to the perspective presented in Sect. 2.1, we can view $\mathbb{SB}$ under the viewpoint of *tropical algebra*, as we show next.

Let S be a commutative semiring and let $I \subseteq S$. We say that I is an *ideal* of S if

$$I + I \subseteq I \quad \text{and} \quad I \cdot S \subseteq I.$$

Let $\mathcal{G}_S = \{a + a \mid a \in S\}$. It is immediate that $\mathcal{G}_S$ is an ideal of S. Let $\nu : S \to \mathcal{G}_S$ be the canonical map defined by $a^\nu = a + a$.

Supertropical (commutative) semirings admit the following axiomatic definition. We say that a commutative semiring S is *supertropical* if, for all $a, b \in S$:

(ST1) $\mathcal{G}_S \subset S$;
(ST2) $(a^\nu)^\nu = a^\nu$;
(ST3) $a + b = a^\nu$ if $a^\nu = b^\nu$;
(ST4) $a + b \in \{a, b\}$ if $a^\nu \neq b^\nu$.

In this case, we say that $\mathcal{G}_S$ is the *ghost ideal* of S and a^ν is the *ghost* of a. The ghost ideal replaces favorably 0 in many instances, namely in the key definition of independence of vectors. In fact, the ghost ideal provides a supertropical semiring with an algebraic and geometric theory much deeper than in the general case of arbitrary (commutative) semirings (see [32]).

© Springer International Publishing Switzerland 2015

J. Rhodes, P.V. Silva, *Boolean Representations of Simplicial Complexes and Matroids*, Springer Monographs in Mathematics,
DOI 10.1007/978-3-319-15114-4

The next result collects some properties that shed some light on the structure of supertropical semirings.

Proposition A.1.1 ([32, Section 3]). *Let S be a supertropical semiring. Then:*

(i) $a + a + a = a + a$ *for every* $a \in S$.
(ii) $(\mathcal{G}_S, +)$ *is a submonoid of* $(S, +)$.
(iii) $\mathcal{G}_S$ *is totally ordered by*

$$a \leq b \quad if\ a + b = b.$$

(iv) *For all* $a, b \in S$,

$$a + b = \begin{cases} a & if\ a^\nu > b^\nu \\ b & if\ a^\nu < b^\nu \\ a^\nu & if\ a^\nu = b^\nu \end{cases}$$

Proof. (i) By (ST2), we have $(a^\nu)^\nu = a^\nu$, hence $a + a^\nu = a^\nu$ by (ST3). That is $a + a + a = a + a$.

(ii) By (ST3) and (ST4), $\mathcal{G}_S$ is closed under addition. Since $0 = 0^\nu$, $(\mathcal{G}_S, +)$ is a submonoid of $(S, +)$.

(iii) If $a \leq b \leq c$ in $\mathcal{G}_S$, then $a + c = a + (b + c) = (a + b) + c = b + c = c$, hence $a \leq c$ and so $\leq$ is transitive. By (ST3) and (ST4), $a + b \in \{a, b\}$ for all $a, b \in \mathcal{G}_S$. Since $+$ is commutative, it follows easily that $\leq$ is a total order on G.

(iv) Let $a, b \in S$. In view of (ST3), we may assume that $a^\nu < b^\nu$. Then $a^\nu + b^\nu = b^\nu$. On the other hand, (ST4) yields $a + b \in \{a, b\}$. Suppose that $a + b = a$. Then

$$b^\nu = a^\nu + b^\nu = a + a + b + b = a + b + a + b = a + a = a^\nu,$$

a contradiction. Thus $a + b = b$ and (iv) holds. $\square$

The next result shows that $\mathbb{SB}$ can be characterized as the smallest supertropical semiring.

Proposition A.1.2 ([28, Appendix B]). $\mathbb{SB}$ *is a supertropical semiring and embeds in every supertropical semiring.*

Proof. It is immediate that $\mathbb{SB}$ satisfies (ST1)–(ST4), hence $\mathbb{SB}$ is a semitropical semiring. Let $(S, +, \cdot, 0, 1)$ be an arbitrary supertropical semiring. We claim that the mapping $\varphi : \mathbb{SB} \to S$ defined by

$$0 \mapsto 0, \quad 1 \mapsto 1, \quad 2 \mapsto 1^\nu$$

is injective. Indeed, if $1 = 1^v$, then $a = a \cdot 1 = a \cdot 1^v$ for every $a \in S$, contradicting (ST1). On the other hand $0 = 1^v$ implies $0 = 1$ by Proposition A.1.1(i), thus φ is injective.

Let $a, b \in \mathbb{SB}$. It remains to be proved that

$$(a + b)\varphi = a\varphi + b\varphi \quad \text{and} \quad (a \cdot b)\varphi = a\varphi \cdot b\varphi.$$

Indeed, Proposition A.1.1(i) implies that $1 + 1^v = 1^v$ holds in S. On the other hand, (ST2) yields $1^v + 1^v = (1^v)^v = 1^v$ and

$$1^v \cdot 1^v = (1 + 1) \cdot 1^v = (1 \cdot 1^v) + (1 \cdot 1^v) = 1^v + 1^v = 1^v.$$

These three equalities imply that $(1 + 2)\varphi = 1\varphi + 2\varphi$, $(2 + 2)\varphi = 2\varphi + 2\varphi$ and $(2 \cdot 2)\varphi = 2\varphi \cdot 2\varphi$, respectively. The remaining cases being immediate, φ is a semiring embedding as claimed.□

Note that $\mathcal{G}_{\mathbb{SB}} = \mathcal{G}$, in the notation introduced in Sect. 2.2.

A general theory of matrices over *supertropical semifields* has been developed by Izhakian and Rowen (see [25, 26, 31–35]) and $\mathbb{SB}$ is just a particular case. In particular, the results in Sect. 2.2 hold in this more general setting.

Tropical algebra has become an important area of research since the tropical context allowed the development of a consistent and rich theory of tropical linear algebra and tropical algebraic geometry (see e.g. [20, 50]).

Interesting refinements have been considered recently, allowing further generalization of important concepts from the classical theory. We refer the reader to the survey article [27] by Izhakian, Knebusch and Rowen.

A.2 Closure Operators and Semilattice Structure

We establish in this section the various equivalent alternatives to the concept of closure operators for lattices.

The following properties can be easily deduced from the axioms.

Lemma A.2.1. *Let* $\xi : L \to L$ *be a closure operator on a lattice* L *and let* $a, b \in L$. *Then:*

(i) $(a \vee b)\xi = (a\xi \vee b\xi)\xi$;
(ii) $(a\xi \wedge b\xi) = (a\xi \wedge b\xi)\xi$;
(iii) $\max\{x \in L \mid x\xi = a\xi\} = a\xi$.

Proof. (i) Recall the axioms (C1)–(C3) from page 20. By (C1), we have $a \leq a\xi$ and $b \leq b\xi$, hence $(a \vee b) \leq (a\xi \vee b\xi)$. Thus $(a \vee b)\xi \leq (a\xi \vee b\xi)\xi$ by (C2).

On the other hand, $a \leq (a \vee b)$ yields $a\xi \leq (a \vee b)\xi$ by (C2). Similarly, $b\xi \leq (a \vee b)\xi$, whence $(a\xi \vee b\xi) \leq (a \vee b)\xi$ and so $(a\xi \vee b\xi)\xi \leq (a \vee b)$ $\xi^2 = (a \vee b)\xi$ by (C2) and (C3).

(ii) We have $(a\xi \wedge b\xi) \leq (a\xi \wedge b\xi)\xi$ by (C1). On the other hand, $(a\xi \wedge b\xi) \leq a\xi$ yields $(a\xi \wedge b\xi)\xi \leq a\xi^2 = a\xi$ by (C2) and (C3). Similarly, $(a\xi \wedge b\xi)\xi \leq b\xi$ and therefore $(a\xi \wedge b\xi)\xi \leq (a\xi \wedge b\xi)$.

(iii) We have $a\xi \in \{x \in L \mid x\xi = a\xi\}$ by (C3). On the other hand, if $x\xi = a\xi$, then $x \leq x\xi = a\xi$ by (C1) and the claim follows.□

The set of all closure operators on L will be denoted by CO L. We define a partial order on CO L by

$$\xi \leq \xi' \quad \text{if } a\xi \leq a\xi' \text{ for every } a \in L.$$

This partial order admits several equivalent formulations, as we show next.

Lemma A.2.2. *Let L be a lattice and let $\xi, \xi' \in$ CO L. Then the following conditions are equivalent:*

(i) $\xi \leq \xi'$;
(ii) $\xi\xi' = \xi'$;
(iii) $\xi'\xi = \xi'$;
(iv) $L\xi \supseteq L\xi'$;
(v) $\operatorname{Ker}\xi \subseteq \operatorname{Ker}\xi'$.

Proof. (i) $\Rightarrow$ (ii). Let $a \in L$. By (C1), we have $a \leq a\xi$, hence $a\xi' \leq a\xi\xi'$ by (C2). On the other hand, $a\xi \leq a\xi'$ yields $a\xi\xi' \leq a\xi'\xi' = a\xi'$ by (C2) and (C3). Thus $\xi\xi' = \xi'$.

(ii) $\Rightarrow$ (iii). Let $a \in L$. By (ii) and (C3), we have $a\xi'\xi\xi' = a\xi'\xi' = a\xi'$, hence $a\xi'\xi \leq a\xi'$ by Lemma A.2.1(iii). Now $a\xi'\xi \geq a\xi'$ follows from (C1).

(iii) $\Rightarrow$ (i). Let $a \in L$. By (C1) and (C2), we have $a\xi \leq a\xi'\xi$. Now (iii) yields $a\xi \leq a\xi'$.

(iii) $\Rightarrow$ (iv). We get $L\xi' = L\xi'\xi \subseteq L\xi$.

(iv) $\Rightarrow$ (iii). Let $a \in L$. Then $a\xi' = b\xi$ for some $b \in L$. Hence $a\xi'\xi = b\xi\xi = b\xi = a\xi'$ by (C3).

(ii) $\Rightarrow$ (v). Let $a, b \in L$ be such that $a\xi = b\xi$. Then $a\xi' = a\xi\xi' = b\xi\xi' = b\xi'$ and so $\operatorname{Ker}\xi \subseteq \operatorname{Ker}\xi'$.

(v) $\Rightarrow$ (ii). Let $a \in L$. Since $(a, a\xi) \in \operatorname{Ker}\xi$ by (C3), we get $(a, a\xi) \in \operatorname{Ker}\xi'$ and so $a\xi' = a\xi\xi'$.□

Lemma A.2.3. *Let L be a lattice. Then $(\text{CO } L, \leq)$ is a lattice and*

$$a(\xi \wedge \xi') = (a\xi \wedge a\xi') \tag{A.1}$$

for all $a \in L$ and $\xi, \xi' \in$ CO L.

Proof. Given $\xi, \xi' \in$ CO L, let $\eta : L \to L$ be defined by

$$a\eta = (a\xi \wedge a\xi')$$

for every $a \in L$. We claim that $\eta \in$ CO L. Indeed, it is immediate that η satisfies (C1) and (C2). It remains to be proved that $a\eta^2 \leq a\eta$.

Since $a\eta \leq a\xi$, we have $a\eta\xi \leq a\xi^2 = a\xi$ by (C2) and (C3). Similarly, $a\eta\xi' \leq a\xi'$ and so

$$a\eta^2 = (a\eta\xi \wedge a\eta\xi') \leq (a\xi \wedge a\xi') = a\eta.$$

Thus $\eta \in CO\,L$.

It is immediate that $\eta \leq \xi, \xi'$. Let $\xi'' \in CO\,L$ be such that $\xi'' \leq \xi, \xi'$. Then, for every $a \in L$, we have $a\xi'' \leq (a\xi \wedge a\xi') = a\eta$ and so $\eta = (\xi \wedge \xi')$. Thus (A.1) holds and $(CO\,L, \leq)$ is a lattice (with the determined join). □

We show next how the semilattice structures of L determined by meet and join relate to closure operators.

We start by considering $\wedge$-subsemilattices. We assume that $\mathrm{Sub}_\wedge L$ is partially ordered by reverse inclusion ($\supseteq$). Since $\mathrm{Sub}_\wedge L$ is closed under intersection, we have

$$(S \vee S') = S \cap S' \tag{A.2}$$

for all $S, S' \in \mathrm{Sub}_\wedge L$ and so $\mathrm{Sub}_\wedge L$ constitutes a lattice (with the determined meet).

The next proposition describes the connection between closure operators and $\wedge$-subsemilattices (see [23, Subsection I.3.12]).

Proposition A.2.4. *Let L be a lattice. Then the mappings*

$$CO\,L \xrightarrow[\Phi']{\Phi} \mathrm{Sub}_\wedge L$$

defined by $\xi\Phi = L\xi$ and $a(S\Phi') = \wedge(S \cap a\uparrow)$ $(a \in L)$ are mutually inverse lattice isomorphisms.

Proof. If $\xi \in CO\,L$, then $T\xi = T$ by (C1). In view of Lemma A.2.1(ii), we get $L\xi \in \mathrm{Sub}_\wedge L$ and so Φ is well defined.

Let $S \in \mathrm{Sub}_\wedge L$ and write $\xi_S = S\Phi'$. It is immediate that ξ_S satisfies axioms (C1) and (C2). To prove (C3), it suffices to show that $S \cap a\uparrow = S \cap a\xi_S\uparrow$, i.e. to prove the equivalence

$$s \geq a \Leftrightarrow s \geq a\xi_S \tag{A.3}$$

for all $a \in L$ and $s \in S$. Indeed, if $s \geq a$, then $s \in S \cap a\uparrow$ and so $s \geq a\xi_S$, and the converse implication follows from (C1). Thus ξ_S satisfies (C3) and so Φ' is well defined.

We show next that $L\xi_S = S$. The direct inclusion follows from $S \in \mathrm{Sub}_\wedge L$. For the opposite inclusion, it suffices to note that $s\xi_S = s$ for every $s \in S$, and this claim follows from (C1) and taking $a = s$ in (A.3). Therefore $\Phi'\Phi$ is the identity mapping on $\mathrm{Sub}_\wedge L$.

This implies that Φ is surjective, and injectivity follows from the equivalence (i) $\Leftrightarrow$ (iv) in Lemma A.2.2. Therefore Φ and Φ' are mutually inverse bijections. By the same equivalence, they are poset isomorphisms, hence lattice isomorphisms.□

The set of all $\vee$-congruences (respectively $\wedge$-congruences) on L will be denoted by $\mathrm{Con}_\vee L$ (respectively $\mathrm{Con}_\wedge L$).

We can assume that $\mathrm{Con}_\vee L$ is partially ordered by inclusion. Since $\mathrm{Con}_\vee L$ is closed under intersection, we have

$$(\sigma \wedge \sigma') = \sigma \cap \sigma' \tag{A.4}$$

for all $\sigma, \sigma' \in \mathrm{Con}_\vee L$ and so $\mathrm{Con}_\vee L$ constitutes a lattice (with the determined join).

The following proposition establishes $\vee$-congruences as kernels of $\vee$-maps.

Proposition A.2.5. *Let σ be an equivalence relation on a lattice L. Then the following conditions are equivalent:*

(i) $\sigma \in \mathrm{Con}_\vee L$;
(ii) $\sigma = \mathrm{Ker}\,\varphi$ for some $\vee$-map of lattices $\varphi : L \to L'$.

Proof. (i) $\Rightarrow$ (ii). Without loss of generality, we may assume that $T\sigma \neq B\sigma$. Let $L' = L/\sigma$ and define a relation $\leq$ on L' by

$$a\sigma \leq b\sigma \quad \text{if } (a \vee b)\sigma b.$$

If $a\sigma a'$ and $b\sigma b'$, then $(a \vee b)\sigma(a \vee b')\sigma(a' \vee b')$ since $\sigma \in \mathrm{Con}_\vee L$, hence the relation is well defined.

Assume that $a\sigma \leq b\sigma \leq c\sigma$. Then $(a \vee b)\sigma b$ yields $(a \vee b \vee c)\sigma(b \vee c)$ and $(b \vee c)\sigma c$ yields $(a \vee b \vee c)\sigma(a \vee c)$. Hence

$$(a \vee c)\sigma(a \vee b \vee c)\sigma(b \vee c)\sigma c$$

and so $a\sigma \leq c\sigma$. Since $\leq$ is clearly reflexive and anti-symmetric, it is a partial order on L'.

Next we prove that the join $a\sigma \vee b\sigma$ exists for all $a, b \in L$. Indeed, we have $a\sigma, b\sigma \leq (a \vee b)\sigma$. Suppose that $a\sigma, b\sigma \leq c\sigma$ for some $c \in L$. Then $(a \vee c)\sigma c$ and $(b \vee c)\sigma c$ yield $(a \vee b \vee c)\sigma(b \vee c)\sigma c$ and so $(a \vee b)\sigma \leq c\sigma$. It follows that

$$(a\sigma \vee b\sigma) = (a \vee b)\sigma$$

and so L' is a lattice (with the determined meet).

It is immediate that the canonical projection

$$\varphi : L \to L'$$
$$a \mapsto a\sigma$$

is a $\vee$-map with kernel σ.

(ii) $\Rightarrow$ (i). Let $a, b, c \in L$. If $a\varphi = b\varphi$, then $(a \lor c)\varphi = (a\varphi \lor c\varphi) = (b\varphi \lor c\varphi) = (b \lor c)\varphi$ and so σ is a $\lor$-congruence. $\square$

We end this section by associating closure operators and $\lor$-congruences, making explicit a construction suggested in [44, Theorem 6.3.7].

Proposition A.2.6. *Let L be a lattice. Then the mappings*

$$\mathrm{CO}\, L \underset{\Psi'}{\overset{\Psi}{\rightleftarrows}} \mathrm{Con}_\lor L$$

defined by $\xi\Psi = Ker\,\xi$ and $a(\sigma\Psi') = \max_L a\sigma$ ($a \in L$) are mutually inverse lattice isomorphisms.

Proof. Let $\xi \in \mathrm{CO}\, L$ and let $a, b, c \in L$ be such that $a\xi = b\xi$. By Lemma A.2.1(i), we have

$$(a \lor c)\xi = (a\xi \lor c\xi)\xi = (b\xi \lor c\xi)\xi = (b \lor c)\xi,$$

hence $Ker\,\xi \in \mathrm{Con}_\lor L$ and so Ψ is well defined.

Now let $\sigma \in \mathrm{Con}_\lor L$ and write $\xi_\sigma = \sigma\Psi'$. Note that $a\xi_\sigma$ is well defined since every class of a $\lor$-congruence must have a maximum element, namely the join of its elements in L. Now axioms (C1) and (C3) follow immediately from $a, a\xi_\sigma \in a\sigma$.

Assume that $a \leq b$ in L. Then $b = (a \lor b)$, hence $(a\xi_\sigma)\sigma a$ yields $(a\xi_\sigma \lor b)\sigma(a \lor b) = b$. It follows that $a\xi_\sigma \leq (a\xi_\sigma \lor b) \leq b\xi_\sigma$, thus (C2) holds and Ψ' is well defined.

Let $\xi \in \mathrm{CO}\, L$ and let $a \in L$. Then

$$a(\xi\Psi\Psi') = \max_L a(\xi\Psi) = \max_L a(Ker\,\xi) = \max_L \{x \in L \mid x\xi = a\xi\}$$
$$= a\xi$$

by Lemma A.2.1(iii), hence $\xi\Psi\Psi' = \xi$.

Now let $\sigma \in \mathrm{Con}_\lor L$ and $a, b \in L$. Then

$$a(\sigma\Psi'\Psi)b \Leftrightarrow a(\sigma\Psi') = b(\sigma\Psi') \Leftrightarrow \max_L a\sigma = \max_L b\sigma \Leftrightarrow a\sigma b,$$

hence $\sigma\Psi'\Psi = \sigma$. Therefore Ψ and Ψ' are mutually inverse bijections. By the equivalence (i) $\Leftrightarrow$ (v) in Lemma A.2.2, they are poset isomorphisms, hence lattice isomorphisms. $\square$

In view of Proposition A.2.5, Proposition A.2.6 establishes also a correspondence between kernels of $\lor$-maps and closure operators.

A.3   Decomposition of ∨-Maps

We show in this section how ∨-maps can be decomposed using the concepts of MPS (introduced in Sect. 3.1) and MPI. We recall also the notation $\rho_{a,b}$ set at the end of Sect. 3.1.

Proposition A.3.1. *Let $\varphi : L \to L'$ be a ∨-surmorphism of lattices. Then:*

(i) *If φ is not one-to-one, then φ factorizes as a composition of MPSs.*
(ii) *If a covers b and b is smi, then $\rho_{a,b}$ is a minimal nontrivial ∨-congruence on L.*
(iii) *φ is an MPS if and only if $\operatorname{Ker}\varphi = \rho_{a,b}$ for some $a, b \in L$ such that a covers b and b is smi.*

Proof. (i) Since L is finite, there exists a minimal nontrivial ∨-congruence $\rho_1 \subseteq \operatorname{Ker}\varphi$ and we can factor φ as a composition $L \to L/\rho_1 \to L'$ (cf. the proof of Proposition A.2.5 in the Appendix). Now we apply the same argument to $L/\rho_1 \to L'$ and successively.

(ii) Let $x \in L$. We must prove that $(x \vee a, x \vee b) \in \rho_{a,b}$. Since b is smi, a is the unique element of L covering b. Hence either $(x \vee b) = b$ or $(x \vee b) \geq a$. In the first case, we get $x \leq b$ and so $(x \vee a) = a$; in the latter case, we get $(x \vee b) = (x \vee (x \vee b)) \geq (x \vee a) \geq (x \vee b)$ and so $(x \vee b) = (x \vee a)$. Hence $(x \vee a, x \vee b) \in \rho_{a,b}$ and so $\rho_{a,b}$ is a (nontrivial) ∨-congruence on L. Minimality is obvious.

(iii) Assume that φ is an MPS and let $a \in L$ be maximal among the elements of L which belong to a nonsingular $\operatorname{Ker}\varphi$ class. Then there exists some $x \in L \setminus \{a\}$ such that $x\varphi = a\varphi$. It follows that $(x \vee a)\varphi = (x\varphi \vee a\varphi) = a\varphi$ and so by maximality of a we get $(x \vee a) = a$ and so $x < a$. Then there exists some $b \geq x$ such that a covers b. Since every ∨-map is order-preserving, we get $a\varphi = x\varphi \leq b\varphi \leq a\varphi$ and so $a\varphi = b\varphi$.

Suppose that b is not smi. Then b is covered by some other element $c \neq a$, hence $b = (a \wedge c)$ and $a, c < (a \vee c)$. It follows that $(a \vee c)\varphi = (a\varphi \vee c\varphi) = (b\varphi \vee c\varphi) = c\varphi$. Since $c \neq (a \vee c)$ and $a < (a \vee c)$, this contradicts the maximality of a. Thus b is smi. Since $\rho_{a,b} \subseteq \operatorname{Ker}\varphi$, it follows from (ii) that $\operatorname{Ker}\varphi = \rho_{a,b}$.

The converse implication is immediate. $\square$

We prove next the dual of Proposition A.3.1 for injective ∨-maps. We say that a ∨-map $\varphi : L \to L'$ is a *maximal proper injective ∨-map* (MPI) of lattices if φ is injective and $L\varphi$ is a maximal proper ∨-subsemilattice of L'. This amounts to saying that φ cannot be factorized as the composition of two proper injective ∨-maps.

Proposition A.3.2. *Let $\varphi : L \to L'$ be an injective ∨-map of lattices. Then:*

(i) *If φ is not onto, then φ factorizes as a composition of MPIs.*
(ii) *If $a \in \operatorname{sji}(L')$, then the inclusion $\iota : L' \setminus \{a\} \to L'$ is an MPI of lattices.*
(iii) *φ is an MPI if and only if $L\varphi = L' \setminus \{a\}$ for some $a \in \operatorname{sji}(L')$.*

Proof. (i) Immediate since L' is finite and each proper injective $\vee$-map increases
   the number of elements.
(ii) Let $x, y \in L' \setminus \{a\}$. Since a is sji, the join of x and y in L' is also the join of
   x and y in $L' \setminus \{a\}$. Hence $L' \setminus \{a\}$ is a $\vee$-semilattice and therefore a lattice
   with the determined meet. Since $(x\iota \vee y\iota) = (x \vee y) = (x \vee y)\iota$, then ι is a
   $\vee$-map. Since $|L' \setminus \operatorname{Im}\iota| = 1$, it must be an MPI.
(iii) Assume that φ is an MPI. Let a be a minimal element of $L' \setminus L\varphi$. We claim that
   a is an sji in L'. Otherwise, by minimality of a, we would have $a = (x\varphi \vee y\varphi)$
   for some $x, y \in L$. Since φ is a $\vee$-map, this would imply $a = (x \vee y)\varphi$,
   contradicting $a \in L' \setminus L\varphi$.

   Thus a is an sji in L' and we can factor $\varphi : L \to L'$ as the composition of
   $\varphi : L \to L' \setminus \{a\}$ with the inclusion $\iota : L' \setminus \{a\} \to L'$. Since φ is an MPI, then
   $\varphi : L \to L' \setminus \{a\}$ must be onto as required.

   The converse implication is immediate.$\square$

Theorem A.3.3. *Let $\varphi : L \to L'$ be a $\vee$-map of lattices. Then φ factorizes as a
composition of MPSs followed by a composition of MPIs.*

Proof. In view of Propositions A.3.1 and A.3.2, it suffices to note that φ can always
be factorized as $\varphi = \varphi_1\varphi_2$ with φ_1 a $\vee$-surmorphism and φ_2 an injective $\vee$-map.
This can be easily achieved taking $\varphi_1 : L \to L\varphi$ defined like φ, and $\varphi_2 : L\varphi \to L'$
to be the inclusion.$\square$

A.4 Lattice Completions of a Poset

We discuss in this section two lattice completions of a poset that have in some sense
dual properties, and how they relate to boolean representability.

Let P be a poset. For every $X \subseteq P$, we write

$$X_d = \cap_{p\in X} \, p\downarrow \, .$$

In particular, $\emptyset_d = P$. Note that X_d is a down set. Moreover, for all $p, p' \in P$, we
have

$$p \leq p' \Leftrightarrow p\downarrow \subseteq p'\downarrow, \quad p < p' \Leftrightarrow p\downarrow \subset p'\downarrow . \tag{A.5}$$

The sets X_d ($X \subseteq P$) are said to be the *flats* of P, and we write

$$\operatorname{Fl} P = \{X_d \mid X \subseteq P\}.$$

Note that, if we write $P' = \{p\downarrow \mid p \in P\} \subseteq 2^P$, then $\operatorname{Fl} P = \widehat{P'}$ and so $(\operatorname{Fl} P, \subseteq)$
constitutes a lattice under intersection and the determined join, called the *lattice of
flats* of P. Recalling the matrix $M(P)$ defined by (3.5) and the notation Z_i from
Sect. 3.4, it follows easily that

$$\mathrm{Fl}\, P \to \mathrm{Fl}\, M(P)$$
$$X_d \mapsto \cap_{p \in X} Z_p$$

is an isomorphism of lattices, since $Z_p = p \downarrow$ for every $p \in P$.

As an example, if P is the poset of (3.6), then the Hasse diagram of $\mathrm{Fl}\, P$ is

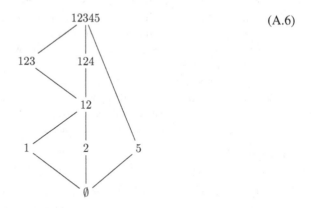

(A.6)

A trivial case arises if our poset is already a lattice.

Proposition A.4.1. *Let L be a lattice. Then*

$$\varphi : L \to \mathrm{Fl}\, L$$
$$a \mapsto a \downarrow$$

is a lattice isomorphism.

Proof. Since $\emptyset_d = T \downarrow$ and $(a \downarrow) \cap (b \downarrow) = (a \wedge b) \downarrow$ for all $a, b \in L$, we have $\mathrm{Fl}\, L = \{a \downarrow \mid a \in L\}$ and so φ is onto. In view of (A.5), φ is injective and an order isomorphism, therefore a lattice isomorphism.□

We say that a mapping $\varphi : P \to P'$ of posets is an *order extension* if $\varphi : P \to \mathrm{Im}\, \varphi$ is an order isomorphism. This is equivalent to saying that

$$p \leq q \Leftrightarrow p\varphi \leq q\varphi$$

holds for all $p, q \in P$. We call a lattice L a *lattice extension* of P if there exists some order extension $\mathcal{P} \to L$. In particular, it follows from (A.5) that $p \mapsto p \downarrow$ defines an order extension of P into $\mathrm{Fl}\, P$, hence $\mathrm{Fl}\, P$ is a lattice extension of P. In some precise sense, it is indeed the "smallest" lattice extension of P, as we show next. Note that $\mathrm{Fl}\, P$ is also known in the literature as the *Dedekind-MacNeille completion* of P (see [3, Section 2.5], [36] and [41, Section 6]).

Proposition A.4.2. *Let $\varphi : P \to L$ be an order extension of a poset P into a lattice L. Then:*

(i) φ induces an order extension $\Phi : \text{Fl } P \to L$;
(ii) If $P\varphi$ $\vee$-generates L, then Φ can be assumed to be a $\wedge$-morphism.

Proof. (i) Clearly,

$$X_d \cap Y_d = (X \cup Y)_d \tag{A.7}$$

holds for all $X, Y \subseteq P$. From the particular case $X_d = Y_d$, it follows that

$$\forall Z \in \text{Fl } P \; \exists Z' \subseteq P \; (Z = Z'_d \wedge \forall X \subseteq P \; (Z = X_d \Rightarrow X \subseteq Z')). \tag{A.8}$$

Indeed, it suffices to take Z' as the union of all the $X \subseteq P$ satisfying $Z = X_d$. We define $\Phi : \text{Fl } P \to L$ by $Z\Phi = \wedge(Z'\varphi)$. For all $X, Y \in \text{Fl } P$, in view of (A.7) and (A.8), we have

$$X \subseteq Y \Leftrightarrow X'_d \subseteq Y'_d \Leftrightarrow X'_d = X'_d \cap Y'_d \Leftrightarrow X'_d = (X' \cup Y')_d$$
$$\Leftrightarrow X' = X' \cup Y' \Leftrightarrow X' \supseteq Y' \Leftrightarrow X'\varphi \supseteq Y'\varphi.$$

Hence $X \subseteq Y$ implies $X\Phi = \wedge(X'\varphi) \leq \wedge(Y'\varphi) = Y\Phi$. Conversely, assume that $X\Phi \leq Y\Phi$. Let $x \in X = X'_d$. Then $x \leq p$ for every $p \in X'$, hence $x\varphi \leq p\varphi$ and so $x\varphi \leq \wedge(X'\varphi) = X\Phi \leq Y\Phi = \wedge(Y'\varphi)$. Thus $x\varphi \leq q\varphi$ for every $q \in Y'$ and so $x \leq q$, yielding $x \in Y'_d = Y$. Therefore $X \subseteq Y$ and so $\Phi : \text{Fl } P \to L$ is an order extension.

(ii) Let $X, Y \in \text{Fl } P$. Since $P\varphi$ $\vee$-generates L, by Lemma 3.3.1(ii) it suffices to show that

$$p\varphi \leq (X\Phi \wedge Y\Phi) \; \Leftrightarrow \; p\varphi \leq (X \cap Y)\Phi \tag{A.9}$$

holds for every $p \in P$. Now

$$(X\Phi \wedge Y\Phi) = (\wedge(X'\varphi)) \wedge (\wedge(Y'\varphi)) = \wedge((X'\varphi) \cup (Y'\varphi))$$
$$= \wedge((X' \cup Y')\varphi).$$

Since $\varphi : \text{Fl } P \to L$ is an order extension, it follows that $p\varphi \leq (X\Phi \wedge Y\Phi)$ if and only if $p\varphi \leq q\varphi$ for every $q \in X' \cup Y'$ if and only if $p \leq q$ for every $q \in X' \cup Y'$. Thus

$$p\varphi \leq (X\Phi \wedge Y\Phi) \Leftrightarrow p \in (X' \cup Y')_d. \tag{A.10}$$

On the other hand, $(X \cap Y)\Phi = \wedge((X \cap Y)'\varphi)$ and we get

$$p\varphi \leq (X \cap Y)\Phi \Leftrightarrow p \in (X \cap Y)'_d. \tag{A.11}$$

Since (A.7) yields $(X' \cup Y')_d = X'_d \cap Y'_d = X \cap Y = (X \cap Y)'_d$, then (A.10) and (A.11) together yield (A.9) and we are done.$\square$

The next examples show that the restrictions in Proposition A.4.2(ii) cannot be omitted. Consider the poset P and its respective lattice of flats described by their Hasse diagrams,

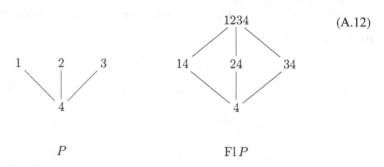

(A.12)

$$P \qquad\qquad \text{Fl}\,P$$

and the following lattices:

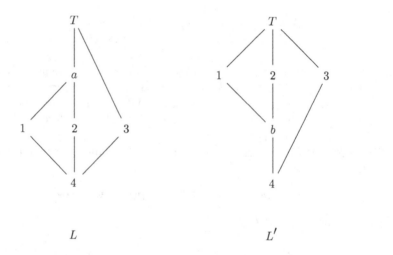

$$L \qquad\qquad L'$$

Let $\varphi : P \to L$ be the inclusion mapping, which is an order extension. Then $P\varphi$ ∨-generates L, but the unique order extension $\Phi : \text{Fl}\,P \to L$ is not a ∨-morphism.

Similarly, let $\varphi' : P \to L'$ be the inclusion mapping, which is also an order extension. It is easy to see that the unique order extension $\Phi' : \text{Fl}\,P \to L'$ is not a ∧-morphism, but of course $P\varphi'$ does not ∨-generates L'.

We intend to take advantage of Proposition 3.6.2 as follows. Given a poset P, we can view $\text{Fl}\,P$ as ∨-generated by P (by identifying $p \in P$ with $p\!\downarrow$). Indeed, since flats are down sets, the equalities

$$X = \cup\{x\!\downarrow \mid x \in X\} = \vee\{x\!\downarrow \mid x \in X\}$$

hold for every $X \in \mathrm{Fl}\,P$ and our claim holds. However, this does not imply that $(\mathrm{Fl}\,P, P) \in \mathrm{FLg}$ since the bottom element of $\mathrm{Fl}\,P$ may be of the form $p \downarrow$ for some $p \in P$, and this happens precisely if P has a minimum. To avoid this situation, we introduce the lattice $\mathrm{Fl}_0\,P$ obtained from $\mathrm{Fl}\,P$ by adding $\emptyset$ as bottom element (if needed). Now we certainly have $(\mathrm{Fl}_0\,P, P) \in \mathrm{FLg}$.

Note also that, since $\mathrm{Fl}_0\,P$ is a $\cap$-subsemilattice of $(2^P, \subseteq)$, it induces a closure operator on 2^P defined by $\mathrm{Cl}_{F_0} = (\mathrm{Fl}_0\,P)\Phi$ by Proposition A.2.4, i.e.

$$\mathrm{Cl}_{F_0} X = \cap \{Z \in \mathrm{Fl}_0\,P \mid X \subseteq Z\}$$

for every $X \subseteq P$.

Proposition A.4.3. *Let P be a poset and $X \subseteq P$. Then the following conditions are equivalent:*

(i) X *is c-independent as a subset of P;*
(ii) X *is c-independent as a subset of $\mathrm{Fl}\,P$;*
(iii) X *admits an enumeration $x_1, \ldots, x_k$ such that*

$$(x_1 \downarrow \vee \ldots \vee x_k \downarrow) \supset \ldots \supset (x_{k-1} \downarrow \vee x_k \downarrow) \supset x_k \downarrow$$

holds in $\mathrm{Fl}\,P$;
(iv) X *admits an enumeration $x_1, \ldots, x_k$ such that*

$$\mathrm{Cl}_{F_0}(x_1, \ldots, x_k) \supset \mathrm{Cl}_{F_0}(x_2, \ldots, x_k) \supset \ldots \supset \mathrm{Cl}_{F_0}(x_k);$$

(v) X *admits an enumeration $x_1, \ldots, x_k$ such that*

$$x_i \notin \mathrm{Cl}_{F_0}(x_{i+1}, \ldots, x_k) \qquad (i = 1, \ldots, k-1);$$

(vi) X *is a transversal of the successive differences for some chain of $\mathrm{Fl}_0\,P$;*
(vii) X *is a partial transversal of the successive differences for some maximal chain of $\mathrm{Fl}_0\,P$.*

Proof. (i) $\Rightarrow$ (ii). Write $M = M(P)$ and $M' = M(\mathrm{Fl}\,P)$. It follows from Proposition 2.2.6 that X is c-independent as a subset of P if and only if there exists some $Y \subseteq P$ such that $M[Y, X]$ is nonsingular. Let $X' = \{x \downarrow \mid x \in X\}$ and $Y' = \{y \downarrow \mid y \in Y\}$. It follows from (A.5) that the matrices $M'[Y', X']$ and $M[Y, X]$ are essentially the same and so X (which is identified with X' in $\mathrm{Fl}\,P$) is c-independent as a subset of $\mathrm{Fl}\,P$.

(ii) $\Rightarrow$ (i). We may assume that there exists some $Z' \subseteq \mathrm{Fl}\,P$ such that $M'[Z', X']$ is nonsingular. Reordering the elements if necessary, we may assume that $M'[Z', X']$ is lower unitriangular. Let the rows (respectively the columns) of $M'[Z', X']$ be ordered by $(Z_1)_d, \ldots, (Z_k)_d$ (respectively $x_1 \downarrow, \ldots, x_k \downarrow$) with $Z_i \subseteq P$ and $x_i \in P$. Then

$$(Z_i)_d \not\supseteq x_i \downarrow \qquad \text{and} \qquad (Z_i)_d \supseteq x_j \downarrow \quad \text{if } j > i. \qquad \text{(A.13)}$$

Since $(Z_i)_d = \cap_{z \in Z_i} z \downarrow$, it follows that, for $i = 1, \ldots, n$, there exists some $z_i \in Z_i$ such that $z_i \downarrow \not\supseteq x_i \downarrow$. Moreover, $z_i \downarrow \supseteq (Z_i)_d \supseteq x_j \downarrow$ whenever $j > i$, hence we may replace $(Z_i)_d$ by $z_i \downarrow$ in (A.13). Hence the matrix $M[Z, X]$ is lower unitriangular and therefore nonsingular for $Z = \{z_1, \ldots, z_k\}$. Thus X is c-independent as a subset of P.

The remaining equivalences from the theorem follow from Proposition 3.6.2 after the following preliminary remarks.

First, we note that the sets $p \downarrow$ are nonempty for all $p \in P$, hence it is indifferent to have $\mathrm{Fl}\, P$ or $\mathrm{Fl}_0\, P$ in (ii) and (iii).

Second, we claim that

$$\mathrm{Fl}(\mathrm{Fl}_0\, P, P) = \mathrm{Fl}_0\, P. \tag{A.14}$$

Indeed, given $X \subseteq P$, it follows from the definition that $X \in \mathrm{Fl}(\mathrm{Fl}_0\, P, P)$ if and only if $X = \{p \in P \mid p \downarrow \subseteq Y\}$ for some $Y \in \mathrm{Fl}_0\, P$. Since $\{p \in P \mid p \downarrow \subseteq Y\} = Y$ due to Y being a down set, (A.14) holds.

Third and last, if $\mathrm{Cl}_L X = \cap \{Y \in \mathrm{Fl}(\mathrm{Fl}_0\, P, P) \mid X \subseteq Y\}$, it follows from (A.14) that $\mathrm{Cl}_L X = \mathrm{Cl}_{F_0} X$.

Now we may safely apply Proposition 3.6.2. $\square$

We can also characterize the c-rank.

Proposition A.4.4. *Let P be a poset. Then* $\mathrm{c} - \mathrm{rk}\, P = \mathrm{ht}\, \mathrm{Fl}_0\, P$.

Proof. We have c-rk $P = \mathrm{rk}\, M(\mathrm{Fl}\, P, P)$ by Proposition A.4.3. If $\mathrm{Fl}_0\, P \neq \mathrm{Fl}\, P$, then P has a minimum and so $M(\mathrm{Fl}_0\, P, P)$ is obtained from $M(\mathrm{Fl}\, P, P)$ by adding an extra row of 1's to a matrix which has already one such row. This operation leaves the rank unchanged, hence c-rk $P = \mathrm{rk}\, M(\mathrm{Fl}_0\, P, P)$ in any case. Since rk $M(\mathrm{Fl}_0\, P, P) = \mathrm{ht}\, \mathrm{Fl}_0\, P$ by Proposition 3.6.4, we get the desired equality. $\square$

For instance, back to our example in (3.6) and (A.6), it follows easily from Hasse $\mathrm{Fl}\, P$ that 5321 is c-independent, by considering the chain

$$12345 \supset 123 \supset 12 \supset 1 \supset \emptyset,$$

which admits 5321 as a transversal for the successive differences. Thus 5321 realizes c-rk $P = \mathrm{ht}\, \mathrm{Fl}_0\, P = 4$.

We associate now another lattice to the poset P. Let Down P be constituted by all the down sets of P, ordered under inclusion.

Lemma A.4.5. *Let P be a poset and let $X \subseteq P$. Then the following conditions are equivalent:*

(i) $X \in \mathrm{Down}\, P$;
(ii) $X = p_1 \downarrow \cup \ldots \cup p_n \downarrow$ *for some* $p_1, \ldots, p_n \in P$;
(iii) $X = (Y_1)_d \cup \ldots \cup (Y_n)_d$ *for some* $Y_1, \ldots, Y_n \subseteq P$.

Proof. (i) $\Rightarrow$ (ii). If $X \in$ Down P, then $X = \cup_{p \in X} p \downarrow$.

(ii) $\Rightarrow$ (iii) and (iii) $\Rightarrow$ (i). Immediate.$\square$

Lemma A.4.6. *Let P be a poset. Then:*

(i) Down P is a sublattice of $(2^P, \subseteq)$;
(ii) $\text{Fl}_0 P \subseteq$ Down P;
(iii) Down P is $\vee$-generated by P (identifying p with $p \downarrow$);
(iv) P is a c-independent subset of Down P.

Proof. (i) Since Down P is clearly closed under union and intersection, and contains both P and $\emptyset$.

(ii) Immediate.

(iii) By Lemma A.4.5(ii).

(iv) By Proposition 3.6.2, it suffices to show that P admits an enumeration $p_1, \ldots, p_k$ such that

$$(p_1 \downarrow \cup \ldots \cup p_k \downarrow) \supset (p_2 \downarrow \cup \ldots \cup p_k \downarrow) \supset \ldots \supset (p_{k-1} \downarrow \cup p_k \downarrow) \supset p_k \downarrow.$$

This can be easily achieved by taking p_1 maximal in P and p_i maximal in $P \setminus \{p_1, \ldots, p_{i-1}\}$ for $i = 2, \ldots, |P|$.$\square$

The following result shows that Down P satisfies in some sense a dual property with respect to Fl P in Proposition A.4.2.

Proposition A.4.7. *Let $\varphi : P \to L$ be an order extension of a poset P into a lattice L $\vee$-generated by $P\varphi$. Then there exists an injective $\wedge$-map $\psi : L \to$ Down P.*

Proof. We define a mapping $\psi : L \to$ Down P by

$$x\psi = \{p \in P \mid p\varphi \leq x\}.$$

Since φ is an order extension, ψ is well defined. Note that $T\psi = P$ and so the top element is preserved.

Let $x, x' \in L$. Then

$$\begin{aligned}
(x \wedge x')\psi &= \{p \in P \mid p\varphi \leq (x \wedge x')\} \\
&= \{p \in P \mid p\varphi \leq x\} \cap \{p \in P \mid p\varphi \leq x'\} \\
&= (x\psi) \cap (x'\psi),
\end{aligned}$$

hence ψ is a $\wedge$-morphism and therefore a $\wedge$-map.

Since L is $\vee$-generated by $P\varphi$, it follows from Lemma 3.3.1(ii) that ψ is injective.$\square$

The next examples show that the restrictions in Proposition A.4.7 cannot be removed. Consider the poset P from (A.12). We compute Down P and define the following lattice L'':

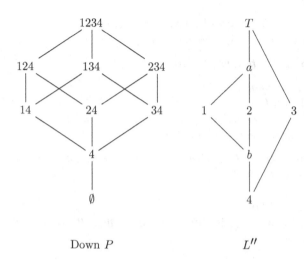

Down P L''

Consider the unique order extension $\varphi : P \to \mathrm{Fl}\, P$ (see (A.12)). It is immediate that $P\varphi$ $\vee$-generates Fl P, but there is no lattice monomorphism $\psi : \mathrm{Fl}\, P \to \mathrm{Down}\, P$.

On the other hand, let $\varphi : P \to L''$ be the inclusion mapping, which is an order extension. Then $P\varphi$ generates L'' as a lattice, but it is not a $\vee$-generating set. In this case, it is easy to see that there is not even an order extension $\psi : L'' \to \mathrm{Down}\, P$.

A.5   Geometric Simplicial Complexes

We present now a brief description of the geometric perspective of simplicial complexes, viewed as subspaces of some Euclidean space $\mathbb{R}^n$. See [24] and [46].

A family of points $X_0, X_1, \ldots, X_k \in \mathbb{R}^n$ is said to be *affinely independent* if the k vectors $X_1 - X_0, \ldots, X_k - X_0$ are linearly independent. The following lemma provides a useful alternative characterization, implying in particular that the definition does not depend on the choice of X_0.

Lemma A.5.1. *Let $X_0, X_1, \ldots, X_k \in \mathbb{R}^n$. Then the following conditions are equivalent:*

(i) $X_0, X_1, \ldots, X_k$ are affinely independent;
(ii) For all $\lambda_i \in \mathbb{R}$,

$$(\sum_{i=0}^{k} \lambda_i X_i = \vec{0} \quad and \quad \sum_{i=0}^{k} \lambda_i = 0) \quad implies \quad \lambda_0 = \ldots = \lambda_k = 0.$$

Proof. (i) $\Rightarrow$ (ii). Assume that $\sum_{i=0}^{k} \lambda_i X_i = \vec{0}$ and $\sum_{i=0}^{k} \lambda_i = 0$. Then

$$\sum_{i=1}^{k} \lambda_i (X_i - X_0) = -(\sum_{i=1}^{k} \lambda_i) X_0 + \sum_{i=1}^{k} \lambda_i X_i = \sum_{i=0}^{k} \lambda_i X_i = \vec{0}$$

and we get $\lambda_1 = \ldots = \lambda_k = 0$ since $X_1 - X_0, \ldots, X_k - X_0$ are linearly independent. Hence also $\lambda_0 = -\sum_{i=1}^{k} \lambda_i = 0$.

(ii) $\Rightarrow$ (i). Suppose that $X_1 - X_0, \ldots, X_k - X_0$ are linearly dependent. Then there exist $\lambda_i \in \mathbb{R}$, not all zero, such that $\sum_{i=1}^{k} \lambda_i (X_i - X_0) = \vec{0}$. Write $\lambda_0 = -\sum_{i=1}^{k} \lambda_i$. Then $\sum_{i=0}^{k} \lambda_i = 0$. Moreover,

$$\sum_{i=0}^{k} \lambda_i X_i = -(\sum_{i=1}^{k} \lambda_i) X_0 + \sum_{i=1}^{k} \lambda_i X_i = \sum_{i=1}^{k} \lambda_i (X_i - X_0) = \vec{0}.$$

Since the λ_i are not all zero, condition (ii) fails.$\square$

Another equivalent formulation of affine independence is the inexistence of an affine subspace of dimension $< k$ in $\mathbb{R}^n$ containing $\{X_0, X_1, \ldots, X_k\}$.

A subspace $S \subseteq \mathbb{R}^n$ is said to be *convex* if, for all $X, Y \in S$, the segment

$$[X, Y] = \{\lambda X + (1 - \lambda) Y \mid 0 \le \lambda \le 1\}$$

is contained in S. The *convex hull* $\langle V \rangle_C$ of a subset $V \subseteq \mathbb{R}^n$ is the smallest convex subset of $\mathbb{R}^n$ containing S.

If $V \subset \mathbb{R}^n$ is a finite nonempty affinely independent set, we call its convex hull $S = \langle V \rangle_C$ a *geometric simplex* in $\mathbb{R}^n$. If $V = \{X_0, X_1, \ldots, X_k\}$, a more constructive description of S is given by

$$S = \{\lambda_0 X_0 + \ldots + \lambda_k X_k \mid \lambda_0, \ldots, \lambda_k \ge 0; \ \lambda_0 + \ldots + \lambda_k = 1\}.$$

The numbers $\lambda_0, \ldots, \lambda_k$ are said to be the *barycentric coordinates* of $X = \lambda_0 X_0 + \ldots + \lambda_k X_k$. It follows easily from Lemma A.5.1 that the barycentric coordinates are unique.

The next result shows that V is actually uniquely determined by S.

Proposition A.5.2. *Let S be a geometric simplex in $\mathbb{R}^n$. Then there exists a unique finite nonempty affinely independent $V \subset \mathbb{R}^n$ such that $S = \langle V \rangle_C$.*

Proof. Write $]Y, Z[= [Y, Z] \setminus \{Y, Z\}$ and let

$$S' = \bigcup_{Y,Z \in S}]Y, Z[.$$

Assume that $S = \langle V \rangle_C$ for some finite nonempty affinely independent $V \subset \mathbb{R}^n$. We claim that

$$V = S \setminus S'. \tag{A.15}$$

Assume that $V = \{X_0, X_1, \ldots, X_k\}$ where all the elements are distinct. Suppose that $X_0 \in S'$. Then $X_0 \in]Y, Z[$ for some $Y = \lambda_0 X_0 + \ldots + \lambda_k X_k$, $Z = \lambda'_0 X_0 + \ldots + \lambda'_k X_k$ with $\lambda_i, \lambda'_i \geq 0$ and $\sum_{i=0}^{k} \lambda_i = \sum_{i=0}^{k} \lambda'_i = 1$. We may write $X_0 = \mu Y + (1 - \mu)Z$ for some $\mu \in]0, 1[$. Hence

$$X_0 = \sum_{i=0}^{k} (\mu \lambda_i + (1 - \mu)\lambda'_i) X_i.$$

Since

$$\sum_{i=0}^{k} (\mu \lambda_i + (1 - \mu)\lambda'_i) = (\mu \sum_{i=0}^{k} \lambda_i) + (1 - \mu) \sum_{i=0}^{k} \lambda'_i = \mu + 1 - \mu = 1,$$

it follows from Lemma A.5.1 that $\mu \lambda_0 + (1 - \mu)\lambda'_0 = 1$ and $\mu \lambda_i + (1 - \mu)\lambda'_i = 0$ for $i \in \{1, \ldots, k\}$. Since $\mu, 1 - \mu > 0$, we get $\lambda_i = \lambda'_i = 0$ for $i \in \{1, \ldots, k\}$. Since $\sum_{i=0}^{k} \lambda_i = \sum_{i=0}^{k} \lambda'_i = 1$, we get $\lambda_0 = \lambda'_0 = 1$ and so $Y = X_0 = Z$, a contradiction. Hence $X_0 \notin S'$. By symmetry, also $X_i \notin S'$ for $i > 0$ and so $V \subseteq S \setminus S'$.

Conversely, let $X \in S \setminus S'$. Then we may write $X = \lambda_0 X_0 + \ldots + \lambda_k X_k$ with $\lambda_i \geq 0$ and $\sum_{i=0}^{k} \lambda_i = 1$. Suppose that $0 < \lambda_0 < 1$ and let

$$Y = \sum_{i=1}^{k} \frac{\lambda_i}{1 - \lambda_0} X_i.$$

Then $\sum_{i=1}^{k} \frac{\lambda_i}{1-\lambda_0} = \frac{1-\lambda_0}{1-\lambda_0} = 1$ and so $Y \in \langle V \rangle_C = S$. Now $X = \lambda_0 X_0 + (1 - \lambda_0)Y \in [X_0, Y]$. In view of Lemma A.5.1, we get $X \neq X_0, Y$, hence $X \in]X_0, Y[$, a contradiction. Thus $\lambda_0 \in \{0, 1\}$ and by symmetry we get $\lambda_i \in \{0, 1\}$ for every i. Therefore $X \in V$ and so (A.15) holds. It follows that V is uniquely determined by S. $\square$

The elements of V are said to be the vertices of $S = \langle V \rangle_C$ and the *dimension* of S is $\dim S = |V| - 1$. A *face* of S is the convex hull of a finite nonempty proper subset of V.

Geometric simplices of dimensions 0,1,2 and 3 are respectively points, segments, triangles and tetrahedra. Beyond these, we get higher dimensional polytopes.

A *geometric simplicial complex* $\mathcal{K}$ in $\mathbb{R}^n$ is a finite nonempty collection of geometric simplices in $\mathbb{R}^n$ such that:

(GS1) Every face of a simplex in $\mathcal{K}$ is in $\mathcal{K}$;

(GS2) The intersection of any two simplices in $\mathcal{K}$ is either empty or a face of both of them.

We show next how the two perspectives, combinatorial and geometric, relate to each other.

Proposition A.5.3. *Let* $\mathcal{K} = \{S_1, \ldots, S_m\}$ *be a geometric simplicial complex in* $\mathbb{R}^n$ *with* $S_i = \langle V_i \rangle_C$. *Let* $V = \bigcup_{i=1}^m V_i$ *and* $H = \{V_1, \ldots, V_m, \emptyset\}$. *Then* $\mathcal{H}_\mathcal{K} = (V, H)$ *is an (abstract) simplicial complex.*

Proof. The claim follows from (GS1) and $\emptyset \in H$.$\quad\square$

Before reversing this correspondence, we note that, for a given geometric simplicial complex $\mathcal{K} = \{S_1, \ldots, S_m\}$ in $\mathbb{R}^n$, the union $\cup \mathcal{K} = S_1 \cup \ldots \cup S_m$ is a subspace of $\mathbb{R}^n$. It has a natural topology as a subspace of $\mathbb{R}^n$ under the usual topology. We say that $\cup \mathcal{K}$ is the *underlying topological space* of $\mathcal{K}$.

Proposition A.5.4. *Let* $\mathcal{H} = (V, H)$ *be an (abstract) simplicial complex with* $H \neq \{\emptyset\}$. *Then:*

(i) There exists some geometric simplicial complex $\mathcal{K}$ *such that* $\mathcal{H}_\mathcal{K} \cong \mathcal{H}$;
(ii) $\cup \mathcal{K}$ *is unique up to homeomorphism.*

Proof. (i) Write $V = \{a_1, \ldots, a_m\}$ and $H \setminus \{\emptyset\} = \{I_1, \ldots, I_k\}$. For every $i \in \{1, \ldots, m\}$, let $X_i \in \mathbb{R}^m$ have the ith coordinate equal to 1 and all the others equal to 0. Write $V' = \{X_1, \ldots, X_m\}$ and define a bijection $\varphi : V \to V'$ by $a_i\varphi = X_i$. For $j \in \{1, \ldots, k\}$, let $S_j = \langle I_j\varphi \rangle_C$ and define $\mathcal{K} = \{S_1, \ldots, S_k\}$.

It follows from Lemma A.5.1 that V' is affinely independent. In particular, each S_i is a simplex in $\mathbb{R}^m$.

It is immediate that $\mathcal{K}$ satisfies (GS1). To prove (GS2), it suffices to show that

$$S_i \cap S_j = \langle (I_i \cap I_j)\varphi \rangle_C \qquad (A.16)$$

holds for all $i, j \in \{1, \ldots, k\}$.

Let $X = (\lambda_1, \ldots, \lambda_m) \in S_i \cap S_j$. Since $X \in S_i$, it follows that $\lambda_r = 0$ if $a_r \notin I_i$. Moreover, $\sum_{p=1}^m \lambda_p = 1$. Similarly, since $X \in S_j$, we have $\lambda_r = 0$ if $a_r \notin I_j$. Thus $\lambda_r = 0$ if $a_r \notin I_i \cap I_j$ and it follows that $X \in \langle (I_i \cap I_j)\varphi \rangle_C$. The opposite inclusion being trivial, (A.16) holds.

Thus $\mathcal{K}$ satisfies (GS2) and is therefore a geometric simplicial complex. Write $\mathcal{H}_\mathcal{K} = (V', H')$ and let $A \subseteq V$. In view of Proposition A.5.2, we have

$$A\varphi \in H' \quad \text{if and only if} \quad \langle A\varphi \rangle_C \in \mathcal{K}. \qquad (A.17)$$

Since V' is affinely independent (even linearly independent), $\langle A\varphi \rangle_C$ univocally determines $A\varphi$ and therefore A. Thus

$$\langle A\varphi \rangle_C \in \mathcal{K} \quad \text{if and only if} \quad A \in H.$$

Together with (A.17), this yields

$$A\varphi \in H' \quad \text{if and only if} \quad A \in H,$$

hence $\mathcal{H}_\mathcal{K} \cong \mathcal{H}$.

(ii) We sketch the proof. Assume now that $\mathcal{H}_{\mathcal{K}} \cong \mathcal{H}_{\mathcal{K}'}$ for some geometric simplicial complexes $\mathcal{K}$ and $\mathcal{K}'$. It is immediate that $\mathcal{K}$ and $\mathcal{K}'$ must have the same number of simplexes, say $\mathcal{K} = \{S_1, \ldots, S_k\}$ and $\mathcal{K}' = \{S'_1, \ldots, S'_k\}$, and there exists some correspondence $S_j \mapsto S'_j$ preserving dimension, faces and intersections. If we denote by $V = \{X_1, \ldots, X_m\}$ (respectively $V' = \{X'_1, \ldots, X'_m\}$) the vertex set of $\mathcal{K}$ (respectively $\mathcal{K}'$), we get a bijection

$$\varphi : V \to V'$$
$$X_i \mapsto X'_i$$

which induces the correspondence $S_j \mapsto S'_j$.

Using the barycentric coordinates, we can now define a mapping $\Phi : \cup \mathcal{K} \to \cup \mathcal{K}'$ by

$$\left(\sum_{i=0}^{m} \lambda_i X_i\right)\Phi = \sum_{i=0}^{m} \lambda_i X'_i.$$

Since the barycentric coordinates are unique and the structures of $\mathcal{K}$ and $\mathcal{K}'$ match, Φ is a bijection. It remains to be seen that it is continuous.

It is immediate that $\Phi|_{S_j} : S_j \to S'_j$ is a homeomorphism for every $j \in \{1, \ldots, k\}$. Let $X \in \cup \mathcal{K}$ and let $\varepsilon > 0$. Since every simplex in $\mathbb{R}_n$ is compact, there exists some open ball $B_{\delta_0}(X)$ in $\mathbb{R}^n$ which intersects only those simplexes containing X, say $S_{j_1}, \ldots, S_{j_p}$. Now, since each $\Phi|_{S_j} : S_j \to S'_j$ is a homeomorphism, there exist $\delta_1, \ldots, \delta_p > 0$ such that

$$\forall Y \in S_{j_q} \ (\ |Y - X| < \delta_q \ \Rightarrow \ |Y\Phi - X\Phi| < \varepsilon \ )$$

holds for $q \in \{1, \ldots, p\}$. Taking $\delta = \min\{\delta_0, \ldots, \delta_p\}$, we obtain

$$\forall Y \in \cup \mathcal{K} \ (\ |Y - X| < \delta \ \Rightarrow \ |Y\Phi - X\Phi| < \varepsilon \ ).$$

Therefore Φ is continuous. By symmetry, it is a homeomorphism.$\square$

Given an (abstract) simplicial complex $\mathcal{H} = (V, H)$ with $H \neq \{\emptyset\}$, a geometric simplicial complex $\mathcal{K}$ satisfying $\mathcal{H}_{\mathcal{K}} \cong \mathcal{H}$ is generically called the *geometric realization* of $\mathcal{H}$ and denoted (up to homeomorphism of the underlying topological space) by $\| \mathcal{H} \|$.

A.6 Rank Functions

Recalling the definition of the rank function r_H in Sect. 6.3, we present now a more abstract viewpoint of rank functions.

Given a function $\varphi : 2^V \to \mathbb{N}$, consider the following axioms for all $X, Y \subseteq V$:

(A1) $X \subseteq Y \Rightarrow X\varphi \leq Y\varphi$;
(A2) $\exists I \subseteq X : |I| = I\varphi = X\varphi$;
(A3) $(X\varphi = |X| \wedge Y \subseteq X) \Rightarrow Y\varphi = |Y|$.

It is easy to see that the three axioms are independent.

Proposition A.6.1. *Given a function $\varphi : 2^V \to \mathbb{N}$, the following conditions are equivalent:*

(i) $\varphi = r_H$ for some simplicial complex $\mathcal{H} = (V, H)$;
(ii) φ satisfies axioms (A1)–(A3).

Proof. (i) $\Rightarrow$ (ii). It follows immediately from the equivalence

$$Xr_H = |X| \Leftrightarrow X \in H.$$

(ii) $\Rightarrow$ (i). Let $H = \{I \subseteq V : I\varphi = |I|\}$. By (A3), H is closed under taking subsets. Taking $X = \emptyset$ in (A2), we get $\emptyset\varphi = 0$, hence $\emptyset \in H$ and so $\mathcal{H} = (V, H)$ is a simplicial complex. Now, for every $X \in V$, we have

$$Xr_H = \max\{|I| \mid I \in 2^X \cap H\} = \max\{|I| \mid I \subseteq X, I\varphi = |I|\}.$$

By (A2), we get $Xr_H \geq X\varphi$, and $Xr_H \leq X\varphi$ follows from (A1). Hence $\varphi = r_H$ as required.$\square$

We collect next some elementary properties of rank functions.

Proposition A.6.2. *Let $\mathcal{H} = (V, H)$ be a simplicial complex and let $X, Y \subseteq V$. Then:*

(i) $Xr_H \leq |X|$;
(ii) $Xr_H + Yr_H \geq (X \cup Y)r_H$;
(iii) $Xr_H + Yr_H \geq (X \cup Y)r_H + (X \cap Y)r_H$ if some maximal $I \in H \cap 2^{X \cap Y}$ can be extended to some maximal $J \in H \cap 2^{X \cup Y}$;
(iv) $Xr_H + Yr_H \geq (X \cup Y)r_H + (X \cap Y)r_H$ if $\mathcal{H}$ is a matroid.

Proof. (i) By (A2).
(ii) Assume that $(X \cup Y)r_H = |I|$ with $I \in H \cap 2^{X \cup Y}$. Then $I \cap X, I \cap Y \in H$ and so

$$(X \cup Y)r_H = |I| \leq |I \cap X| + |I \cap Y| \leq Xr_H + Yr_H.$$

(iii) We may assume that $(X \cup Y)r_H = |J|$ and $(X \cap Y)r_H = |J \cap X \cap Y|$. It follows that

$$(X \cup Y)r_H + (X \cap Y)r_H = |J| + |J \cap X \cap Y| = |J \cap X| + |J \cap Y|$$
$$\leq Xr_H + Yr_H.$$

(iv) This is well known, but we can include a short deduction from (iii) for completeness.

Let $A \subseteq B \subseteq V$, and assume that $I \in H \cap 2^A$ is maximal. Let $J \in H \cap 2^B$ be maximal and contain I. It follows from (EP') that $|J| = Br_H$. Now we apply part (iii) to $A = X \cap Y$ and $B = X \cup Y$.□

In the next result, we apply the rank function to flats.

Proposition A.6.3. *Let $\mathcal{H} = (V, H)$ be a simplicial complex of rank r.*

(i) If $X, Y \in \mathrm{Fl}\, \mathcal{H}$ and $X r_H = Y r_H$, then

$$X \subseteq Y \quad \textit{if and only if} \quad X = Y.$$

(ii) V *is the unique flat of rank r.*

Proof. (i) Assume that $X \subseteq Y$ and let $I \in H \cap 2^X$ satisfy $|I| = X r_H = Y r_H$. If $p \in Y \setminus X$, then X closed yields $I \cup \{p\} \in H$ and $Y r_H > |I| = X r_H$, a contradiction. Therefore $X = Y$ and (i) holds.

(ii) By part (i).□

It follows that the flats of rank $r - 1$ are maximal in $\mathrm{Fl}\, \mathcal{H} \setminus \{V\}$. Such flats are called *hyperplanes*.

The following result relates the rank function with the closure operator Cl induced by a simplicial complex.

Proposition A.6.4. *Let $\mathcal{H} = (V, H)$ be a boolean representable simplicial complex and let $X \subseteq V$. Then $X r_H$ is the maximum k such that*

$$\mathrm{Cl}(x_1, \ldots, x_k) \supset \mathrm{Cl}(x_2, \ldots, x_k) \supset \ldots \supset \mathrm{Cl}(x_k) \supset \mathrm{Cl}(\emptyset)$$

holds for some $x_1, \ldots, x_k \in X$.

Proof. It follows from Theorem 5.2.6 and the definition of r_H.□

Bibliography

1. Assmus, E.F., Jr., Key, J.D.: Designs and Their Codes. Cambridge University Press, Cambridge (1994)
2. Berstel, J., Perrin, D., Reutenauer, C.: Codes and Automata. Encyclopedia of Mathematics and Its Applications, vol. 129. Cambridge University Press, Cambridge (2010)
3. Birkhoff, G.: Lattice Theory. AMS Colloquium Publications, vol. 25, 3rd edn. American Mathematical Society, Providence (1973)
4. Björner, A.: Shellable and Cohen-Macaulay partially ordered sets. Trans. Am. Math. Soc. **260**, 159–183 (1980)
5. Björner, A., Wachs, M.L.: Nonpure shellable complexes and posets I. Trans. Am. Math. Soc. **348**, 1299–1327 (1996)
6. Björner, A., Wachs, M.L.: Nonpure shellable complexes and posets II. Trans. Am. Math. Soc. **349**, 3945–3975 (1997)
7. Björner, A., Wachs, M., Welker, V.: On sequentially Cohen-Macaulay complexes and posets. Isr. J. Math. **169**, 295–316 (2009)
8. Björner, A., Ziegler, G.M.: Introduction to Greedoids. In: White, N. (ed.) Matroid Applications. Encyclopedia of Mathematics and Its Applications, vol. 40, pp. 284–357. Cambridge University Press, Cambridge/New York (1992)
9. Borovik, A.V., Gelfand, I.M., White, N.: On exchange properties for coxeter matroids and oriented matroids. Discret. Math. **179**(1–3), 59–72 (1998)
10. Brouwer, A.E., Haemers, W.H., Tonchev, V.D.: Embedding partial geometries in Steiner designs. In: Hirschfeld, J.W.P., Magliveras, S.S., de Resmini, M.J. (eds.) Geometry, Combinatorial Designs and Related Structures, pp. 33–41. Proceedings of the First Pythagorean Conference. Cambridge University Press, Cambridge/New York (1997)
11. Cameron, P.J., Gadouleau, M., Mitchell, J.D., Peresse, Y.: Chains of subsemigroups (2015, preprint). arXiv:1501.06394
12. Clifford, A.H., Preston, G.B.: The Algebraic Theory of Semigroups. Mathematical Surveys of the American Mathematical Society, vol. 1, No. 7. American Mathematical Society, Providence (1961)
13. Coxeter, H.S.M.: Self-dual configurations and regular graphs. Bull. Am. Math. Soc. **56**, 413–455 (1950)
14. De Clerck, F., Maldeghem, H.V.: Some classes of rank 2 geometries. In: Buekenhout, F. (ed.) Handbook of Incidence Geometry, Buildings and Foundations, Chapter 10, pp. 433–475. North-Holland, Amsterdam (1995)

© Springer International Publishing Switzerland 2015

J. Rhodes, P.V. Silva, *Boolean Representations of Simplicial Complexes and Matroids*, Springer Monographs in Mathematics, DOI 10.1007/978-3-319-15114-4

15. Develin, M., Santos, F., Sturmfels, B.: On the rank of a tropical matrix. In: Goodman, J.E., Pach, J., Welzl, E. (eds.) Discrete and Computational Geometry. MSRI Publications, vol. 52, pp. 213–242. Cambridge University Press, Cambridge (2005)

16. Diestel, R.: Graph Theory. Springer, New York (2000)

17. Doubilet, P., Rota, G.-C.: On the foundations of combinatorial theory VI: the idea of generating function. In: Proceedings of the sixth Berkeley simposium on mathematical statistics and probability, Berkeley, 1970/71. Volume II: Probability Theory, pp. 267–318. University of California Press, Berkeley (1972)

18. Dress, A.W.M., Wenzel, W.: Endliche Matroide mit Koeffizienten. Bayreuther Mathematische Schriften **26**, 37–98 (1998)

19. Duval, A.M.: Algebraic shifting and sequentially Cohen-Macaulay simplicial complexes. Electron. J. Comb. **3**(1) (1996)

20. Gathmann, A.: Tropical algebraic geometry. Jahresber. Deutsch. Math.-Verein. **108**(1), 3–32 (2006)

21. Geelen, J., Gerards, S., Whittle, G.: Structure in minor-closed classes of matroids. In: Blackburn, S., Gerke, S., Wildon, M. (eds.) Surveys in Combinatorics 2013. London Mathematical Society Lecture Notes Series, vol. 409, pp. 327–362. Cambridge University Press, Cambridge (2013)

22. Gierz, G., Hofmann, K.H., Keimel, K., Lawson, J.D., Mislove, M., Scott, D.S.: Continuous Lattices and Domains. Volume 93 of Encyclopedia of Mathematics and Its Applications. Cambridge University Press, Cambridge (2003)

23. Grätzer, G.: Lattice Theory: Foundation. Springer, Basel (2011)

24. Hatcher, A.: Algebraic Topology. Cambridge University Press, Cambridge/New York (2002)

25. Izhakian, Z.: The tropical rank of a tropical matrix (2006, preprint). arXiv:math.AC/0604208

26. Izhakian, Z.: Tropical arithmetic and tropical matrix algebra. Commun. Algebra **37**(4), 1–24 (2009)

27. Izhakian, Z., Knebusch, M., Rowen, L.: Algebraic structures of tropical mathematics. Contemp. Math. **616**, 125–150 (2014)

28. Izhakian, Z., Rhodes, J.: New representations of matroids and generalizations (2011, preprint). arXiv:1103.0503

29. Izhakian, Z., Rhodes, J.: Boolean representations of matroids and lattices (2011, preprint). arXiv:1108.1473

30. Izhakian, Z., Rhodes, J.: C-independence and c-rank of posets and lattices (2011, preprint). arXiv:1110.3553

31. Izhakian, Z., Rowen, L.: The tropical rank of a tropical matrix. Commun. Algebra **37**(11), 3912–3927 (2009)

32. Izhakian, Z., Rowen, L.: Supertropical algebra. Adv. Math. **225**(8), 2222–2286 (2010)

33. Izhakian, Z., Rowen, L.: Supertropical matrix algebra. Isr. J. Math. **182**, 383–424 (2011)

34. Izhakian, Z., Rowen, L.: Supertropical matrix algebra II: solving tropical equations. Isr. J. Math. **186**(1), 69–97 (2011)

35. Izhakian, Z., Rowen, L.: Supertropical matrix algebra III: powers of matrices and their supertropical eigenvalues. J. Algebra **341**(1), 125–149 (2011)

36. Lascoux, A., Schützenberger, M.-P.: Treillis et bases des groupes de Coxeter. Electron. J. Comb. **3**, #R27 (1996)

37. Lyndon, R.C., Schupp, P.E.: Combinatorial Group Theory. Springer, Berlin/New York (1977)

38. Margolis, S.W., Dinitz, J.H.: Translational hulls and block designs. Semigroup Forum **27**, 247–263 (1983)

39. Oxley, J.G.: Matroid Theory. Oxford Science Publications. Oxford University Press, Oxford/New York (1992)

40. Oxley, J.G.: What is a matroid? Cubo **5**, 179–218 (2003)

41. Reading, N.: Order dimension, strong Bruhat order and lattice properties for posets. Order **19**(1), 73–100 (2002)

42. Rhodes, J., Silva, P.V.: A new notion of vertex independence and rank for finite graphs. Int. J. Algebra Comput. (2012). arXiv:1201.3984. doi:10.1142/S021819671540007X

43. Rhodes, J., Silva, P.V.: Matroids, hereditary collections and simplicial complexes having boolean representations (2012, preprint). arXiv:1210.7064
44. Rhodes, J., Steinberg, B.: The q-Theory of Finite Semigroups. Springer Monographs in Mathematics. Springer, New York (2009)
45. Rota, G.-C.: On the foundations of combinatorial theory I: theory of Möbius functions. Z. Wahrsch. Verw. Gebiete 2, 340–368 (1964)
46. Rotman, J.J.: An Introduction to Algebraic Topology. Springer, New York (1986)
47. Ryser, H.J.: An extension of a theorem of de Bruijn and Erdös on combinatorial designs. J. Algebra 10(2), 246–261 (1968)
48. Ryser, H.J.: Subsets of a finite set that intersect each other in at most an element. J. Comb. Theory (A) 17, 59–77 (1974)
49. Semple, C., Whittle, G.: Partial fields and matroid representation. Adv. Appl. Math. 17(2), 184–208 (1996)
50. Speyer, D.E., Sturmfels, B.: Tropical mathematics. Math. Mag. 82, 163–173 (2009)
51. Stanley, R.P.: Combinatorics and Commutative Algebra, 2nd edn. Birkhäuser, Boston (1995)
52. Stanley, R.P.: An introduction to hyperplane arrangements. In: Miller, E., Reiner, V., Sturmfels, B. (eds.) Geometric Combinatorics. IAS/Park City Mathematics Series, vol. 13, pp. 389–496. American Mathematical Society, Providence (2007)
53. Wachs, M.: Poset topology: tools and applications. Geom. Comb. 13, 497–615 (2007)
54. White, N. (ed.): Theory of Matroids. Encyclopedia of Mathematics and Its Applications, vol. 26. Cambridge University Press, Cambridge/New York (1986)
55. Whitney, H.: On the abstract properties of linear dependence. Am. J. Math. 57(3), 509–533 (1935) (The Johns Hopkins University Press). (Reprinted in Kung (1986), pp. 55–79)

Notation Index

© Springer International Publishing Switzerland 2015
J. Rhodes, P.V. Silva, *Boolean Representations of Simplicial Complexes and Matroids*, Springer Monographs in Mathematics,
DOI 10.1007/978-3-319-15114-4

General Index

© Springer International Publishing Switzerland 2015
J. Rhodes, P.V. Silva, *Boolean Representations of Simplicial Complexes
and Matroids*, Springer Monographs in Mathematics,
DOI 10.1007/978-3-319-15114-4

Printed in the United States
By Bookmasters